Praise for *Crossing the Energy Divide*

"In a period when concerns about global energy security and reducing our emissions of greenhouse gasses dominate public attitudes and policy initiatives, this book by Bob and Ed Ayres comes as a remarkable piece of knowledge, based on insightful research. Readers among the public at large and policymakers would find this an excellent reference volume in defining the future of energy across the globe."

> **Dr. R. K. Pachauri**, Director General, The Energy and Resources Institute (TERI); Chairman, Intergovernmental Panel on Climate Change (IPCC)

"Most observers of U.S. energy policy might think of energy efficiency as a useful tool to manage the growth of energy consumption. They might also see it as a means to ease our transition into a post-carbon world. And, yes, the evidence does support both of these notions. But as Bob and Ed Ayres convincingly document in this important new book, there is much, much more to the energy story. Indeed, their analysis compels a significantly greater attention to the critical role of energy productivity in maintaining a robust economy. In short, energy efficiency (more correctly, exergy efficiency) should be seen as a critical economic resource—one that demands a more informed dialogue and action, a greater level of innovation, and an accelerated investment if we take seriously our responsibility to the future."

> **John A. "Skip" Laitner**, Director, Economic and Social Analysis, American Council for an Energy-Efficient Program

"It's exciting to see some of the riddles of economic history solved. Energy efficiency may become the master key to future wealth!"

> **Ernst von Weizsacker**, Dean, Bren School of Environmental Science and Management, University of California, Santa Barbara

"This book makes coherent and rigorous arguments that increasing energy efficiency is the primary driver of economic growth today and is key to managing climate change."

> **Kandeh Yumkella**, Director-General, United Nations Industrial Development Organization (UNIDO)

"*Crossing the Energy Divide* has appeared right when those of us in the energy sector need it most. Robert and Edward Ayres do an excellent job of explaining that every product and service has an energy component, and the cost of that component will determine our future quality of life. The energy component is now as important as capital and labor in economic modeling. Aside from explaining the importance of new economic models putting energy in its rightful place, the authors go on to address the subjects of renewable energy viability, future decline of traditional fossil fuels, economics of bio-fuels, and the advantages of co-generation. They have produced a viable and well-thought-out roadmap for our entry into a workable energy-based future."

John K. Cool, P.E., C.P.E., President, PowerPlus Engineering, Inc.

"This book is 'mythbusters' applied to the conventional wisdom that reducing greenhouse gas emissions will also reduce income. The counter-conventional wisdom premises are simple: 1) The real driver of economic growth is access to energy services, 2) We convert only 13% of the potential work in our primary energy to useful energy services, and 3) Existing, proven technology can double this efficiency and save money, thus creating a bridge to a nonfossil future. Rich discussions explain each premise and explain how policy changes could make efficiency the fuel of the future. Anyone interested in preserving planet earth without destroying the economy will revel in the deep insights. The authors show how inducing greater fossil efficiency will both mitigate climate change and preserve/grow societal income. The work should switch the debate from what it will cost to mitigate climate change to a search for policy changes to spur profitable efficiency investments in generation and use of energy services."

Thomas R. Casten, Chair, Recycled Energy Development, LLC

Crossing the Energy Divide

Crossing the Energy Divide

Moving from Fossil Fuel Dependence to a Clean-Energy Future

Robert U. Ayres
Edward H. Ayres

Vice President, Publisher: Tim Moore
Associate Publisher and Director of Marketing: Amy Neidlinger
Wharton Editor: Steve Kobrin
Executive Editor: Jeanne Glasser
Editorial Assistant: Myesha Graham
Development Editor: Russ Hall
Operations Manager: Gina Kanouse
Senior Marketing Manager: Julie Phifer
Publicity Manager: Laura Czaja
Assistant Marketing Manager: Megan Colvin
Cover Designer: Chuti Prasertsith
Managing Editor: Kristy Hart
Senior Project Editor: Lori Lyons
Copy Editor: Krista Hansing Editorial Services, Inc.
Proofreader: Kay Hoskin
Indexer: Cheryl Lenser, Ken Johnson
Senior Compositor: Jake McFarland
Manufacturing Buyer: Dan Uhrig

© 2010 by Pearson Education, Inc.
Publishing as Wharton School Publishing
Upper Saddle River, New Jersey 07458

Wharton School Publishing offers excellent discounts on this book when ordered in quantity for bulk purchases or special sales. For more information, please contact U.S. Corporate and Government Sales, 1-800-382-3419, corpsales@pearsontechgroup.com. For sales outside the U.S., please contact International Sales at international@pearson.com.

Company and product names mentioned herein are the trademarks or registered trademarks of their respective owners.

Printed in the United States of America

First Printing December 2009

ISBN-10 0-13-701544-5
ISBN-13 978-0-13-701544-3

Pearson Education LTD.
Pearson Education Australia PTY, Limited.
Pearson Education Singapore, Pte. Ltd.
Pearson Education North Asia, Ltd.
Pearson Education Canada, Ltd.
Pearson Educación de Mexico, S.A. de C.V.
Pearson Education—Japan
Pearson Education Malaysia, Pte. Ltd.

Library of Congress Cataloging-in-Publication Data

Ayres, Robert U.

   Crossing the energy divide : moving from fossil fuel dependence to a clean-energy future / Robert U. Ayres, Edward H. Ayres.

      p. cm.

   Includes bibliographical references and index.

   ISBN 978-0-13-701544-3 (hbk. : alk. paper) 1. Energy development—Economic aspects—United States. 2. Renewable energy sources—United States. 3. Energy consumption—United States. 4. Energy policy—United States. I. Ayres, Ed. II. Title.

   HD9502.U52A97 2010

   333.790973—dc22

                              2009034640

To Leslie and Sharon, our wives.

Contents

Acknowledgments

The list of colleagues, past and present, from whom we have learned, is far too long to fit into a page. People who have inspired our work include Kenneth Boulding, Lester Brown, Colin Campbell, Al Gore, Allen Kneese, Jean LaHerriere, Amory Lovins, Robert Repetto, the old gang at RFF, Vaclav Smil, James Gustave Speth, and Ernst von Weizsaecker. People who have contributed advice, criticism, or information more directly, one way or another, include (alphabetically) Kenneth Arrow, Leslie Ayres, Christian Azar, Thomas Casten, Paul David, Nina Eisenmenger, Arnulf Gruebler, Jean-Charles Hourcade, Marina Fischer-Kowalski, Astrid Kander, Paul Kleindorfer, Arkady Kryazhimsky, Reiner Kuemmel, Jie Li, Skip Luken, Katalin Martinas, Shunsuke Mori, Neboysa Nakicenovich, Tom Prugh, Donald Rogich, Adam Rose, Warren Sanderson, Jerry Silverberg, Thomas Sterner, David Strahan, Jeroen van den Bergh, Benjamin Warr, and Chihiro Watanabe. We also thank the European Commission, INSEAD, IIASA, UN University, and Worldwatch Institute for their support of our work at various times during the past ten years. Finally, we want to thank our publisher, Tim Moore, and our editor, Jeanne Glasser, for their unhesitating willingness to air a very daring challenge to the dominant economic paradigm. Any errors are entirely our responsibility.

About the Authors

Robert U. Ayres is a physicist and economist noted for his work on the role of thermodynamics in the economic process, and more recently for his investigation of the role of energy in economic growth. He is Emeritus Professor of Economics and Technology at the international business school INSEAD, in France, where he has continued his lifelong, pioneering studies of materials/energy flows in the global economy. He originated the concept of industrial metabolism, which has since become a field of study explored by the *Journal of Industrial Ecology*.

Ayres was trained as a physicist at the University of Chicago, University of Maryland, and Kings College London (Ph.D. in Mathematical Physics). He was Professor of Engineering and Public Policy at Carnegie-Mellon University in Pittsburgh from 1979 until 1992, when he was appointed Professor of Environment and Management at INSEAD. He is also an Institute Scholar at the International Institute for Applied Systems Analysis (IIASA) in Austria.

Robert Ayres is author or coauthor of 18 books and more than 200 journal articles and book chapters. His books range from *Alternatives to the Internal Combustion Engine*, with Richard A. McKenna (Johns Hopkins Press, 1972), to *Turning Point: The End of the Growth Paradigm* (Earthscan, 1998) to *The Economic Growth Engine: How Energy and Work Drive Material Prosperity*, with Benjamin Warr (Edward Elgar, 2009). He and his wife reside in Paris.

Edward (Ed) Ayres was Editorial Director at the Worldwatch Institute in Washington, D.C. (publisher of the annual *State of the World* and bi-annual *Vital Signs*) from 1994 through 2005. He also served as editor of the bimonthly *World Watch* magazine during this period. *World Watch* articles and essays by Ayres were distributed to the global media by the Los Angeles Times Syndicate. His writing has also appeared in *Time* magazine in its series "Beyond 2000: Your Health, Our Planet"; *Utne Reader*; *The Ecologist*; and other publications.

Ayres has pursued a lifelong interest in the relationships between individual human health and endurance and the sustainability of human societies. He was the third-place finisher in the first New York Marathon in 1970, and today continues to write and run long distances in the mountains of California, where he and his wife have built an eco-friendly house.

Introduction: The Chasm to Be Crossed

This book makes two paradigm-challenging claims.

First, physical energy plays a far more fundamental role in economic productivity and growth than most of the economists advising business and government have ever acknowledged. The implications for everyone who breathes, especially during the coming period of hoped-for recovery and transition to the clean-energy economy of the future, are enormous. Energy services aren't just a large part of the economy; they're a major part of what *drives* the economy. And if that is so, both the economic recovery and the energy transition will take far longer than the Obama administration has counted on—*unless* investment is targeted to the very specific technologies and industries that make energy services cheaper. Shotgun spending won't do that.

Second, the energy economy of the industrial world is so deeply dependent on fossil fuels that even the fastest conceivable growth of wind, solar, and other renewable-energy industries cannot substantially replace oil, coal, and natural gas for at least several decades. Virtually the entire capital infrastructure of the country—roads and highways, electric power plants, transmission lines, airlines, shipping, steel, chemicals, construction, and home heating and cooling—depends on fossil fuels. Even if the use of electric cars and solar roof panels were to grow as fast as the Internet did, they would still account for only a drop in the ocean of energy we will use during the next two decades.

Alternative energy has begun to make inroads at the margins, but it has much farther to go than those who have called for a "green energy revolution" have acknowledged. In 2007, after two decades of increasingly urgent warnings by climate scientists that we must sharply reduce carbon emissions, and after more than a decade of seemingly accelerating progress toward a greener future, the total percentages of U.S. electric power produced by renewables (other than hydroelectric power, which can no longer increase) were as follows:

Biofuels	1.4
Wind	0.8
Geothermal	0.4
Solar (photovoltaic)	0.1

Recent progress in these industries has been dramatic. However, even with a crash effort comparable to the U.S. mobilization for World War II, or the Apollo program to put a man on the moon, it will take decades for these new energy industries to reach the necessary scale.

And what happens between now and then? The brutal answer is that if the United States were simply to shift most of its energy and climate attention to that long-term sustainable future we now envision, the existing energy economy would likely collapse before the country could reach that future, as surely as a heart-transplant patient would die if a new heart were not available in time.

There may seem to be a large disconnect between this perspective and the widely publicized claim of Al Gore and others that the United States can harness renewables to achieve full energy independence within ten years. Although we share Gore's sense of urgency about the need to replace fossil fuels as quickly as humanly possible, a sober review of the science and economics makes it clear that it could take half a century before the United States fully achieves this goal.

There is a logical solution to this conundrum, however, and it is the great fortune of the United States—and the world—that a viable means of achieving this solution is still within our grasp. It is not a solution that has anywhere near the idealistic appeal of the alternative-energy vision that has begun to galvanize our most progressive leaders, but it is essential to achieving what they envision.

The solution is to *radically reform our management of the existing, fossil fuel–based system* so that we essentially double the amount of energy service we get from each barrel of oil (or "oil-equivalent" of coal or natural gas) during the years it takes to bring carbon-free renewables to the point at which they can truly begin to take over. This is not to echo the heroically hopeful stance of a John F. Kennedy calling for landing a man on the moon. It is not one of those epic goals that we can achieve only by a massive mobilization of technological research and development. It is not an echo of writer Thomas Friedman's call to "get 10,000 inventors working in 10,000 companies and 10,000 garages and 10,000 laboratories to drive transformational breakthroughs," which is a terrific pipe dream of free-market ideology but at best will take a generation to pay off. To safely cross the economic chasm we now face, we need a solution that yields results more quickly. As it happens, the means to a rapid doubling of U.S. energy service (the amount of useful work done by each unit of fossil fuel burned) already exist. Some of these means are hidden from public view and have not been discussed in mainstream media, but are already being profitably used by hundreds of companies and institutions. They *could* be used by tens of thousands more.

We can appreciate how the confluence of high-profile events around the time of the Obama election brought a new sense of excitement and possibility to people who had been discouraged for years. The Phoenix-like return of Al Gore with his galvanizing film about climate change, *An Inconvenient Truth*; Tom Friedman's rallying call for a more muscular "green revolution"; and the remarkable fact that *both* presidential candidates in 2008 recognized the coming threat of global warming and the need for investment in alternative energy—it all generated tremendous eagerness to leap ahead to the clean-energy economy of the future. For nearly two decades, progressive Americans had felt dismissed and ignored at every turn: the Senate voting 95–0 to reject the Kyoto climate treaty in 1995; the second Bush administration refusing to acknowledge that global warming was real (or, later, that it was driven by human activity); vice president Dick Cheney scorning energy efficiency as a feel-good thing that did nothing for the nation's "real" energy needs; and the Iraq War being widely suspected of being mainly an oil war. For Americans who had felt gloom for half their lives, the election of a candidate who had opposed that war and who was a strong advocate of wind and solar investment seemed like a time to celebrate.

That flush of optimism soon faded as the economy continued to deteriorate in 2009. Yet the belief that big investments in renewables could help spur economic recovery remained unquestioned. In the rush of the new administration and Congress to stop the economic hemorrhage, virtually no thought was given to the possibility that a very different kind of climate and energy management might be needed, not only to bridge the chasm that separates today's world from the clean-energy economy of the future, but also to restore enough growth to ensure that the nation can reach the far side of that chasm. The country isn't ready—because the technology and infrastructure aren't ready—for us to jump with both feet onto the wind, solar, and biofuels bandwagon quite yet. There is critical business to take care of first.

For those who feel a great relief and reassurance about what now appears to be an open path to the renewable-energy future, here is an example of the kind of initially disillusioning—but ultimately liberating—assessment that is essential to taking that path safely. A few years ago, behind the gates of a large rust-belt factory in Indiana, the world's largest steel company, Mittal Steel (now Arcelor Mittal), began operating a facility that captured waste heat from one of its fossil fuel–burning processes and converted that heat to emissions-free electricity. A few miles down the road, a rival company, U.S. Steel, was using a similar strategy to generate emissions-free power from waste blast-furnace gas. In 2005, the two rust-belt rivals generated a combined 190 megawatts (MW) of carbon-free energy from their waste—*more than the entire U.S. production of solar-photovoltaic electricity that year.* That was just the waste heat from two fossil fuel–burning plants in one corner of one state.

The production of photovoltaic power has continued to grow rapidly since then. In January 2009, the California company Sempra Energy began operating a 10MW solar farm in Nevada and generating its power at a surprisingly competitive price. Another California company, BrightSource Energy, announced in 2009 that it would build a 100MW thermal solar plant in the Mojave Desert, with construction to be completed by 2013. Solar power will continue to grow dramatically, as will wind power and other carbon-free sources of energy. But renewables have started from such a tiny base (solar and wind energy together produced less than 1 percent of U.S. electric power in 2007) that, even with geometric growth, they will need 20 years or more before they can replace a big share of the millions of coal-, oil-, and natural gas–burning steam generators, factories, and engines that power our civilization—and our economy. Meanwhile, the near-term potential for ramping up the clean-energy supply by such already-proven means as the ones being exploited by the Mittal and U.S. Steel plants is far larger. For every one of the roughly 1,000 American industrial plants doing this kind of waste-energy recycling, 10 more have yet to begin. Environmentalists might be disheartened and disoriented to think that the fastest and cheapest way to cut carbon emissions and fossil fuel use is *not* to turn our backs on the dirty old industries of the past and present, but to wade into their most neglected corners and clean them up, until the more ideal alternatives are up to scale.

By using the phrase "clean them up," we're not alluding to so-called "clean-coal" processing or to elaborate schemes for capturing carbon emissions and pumping the carbon deep into the ground or the ocean. Those kinds of would-be cleanup are prohibitively costly and even less ready for prime time than solar power. And if the technology for carbon capture and sequestration eventually makes economic sense, the facilities will likely take many years to build—not a realistic option during a time when the country has entered economic survival mode. The strategy we describe in this book doesn't depend on yet-to-be-developed, we-think-it-will-work technologies. Instead, it's an *energy-management* strategy, entailing a sweeping reassessment of ideological blind spots, structural barriers, bad habits, and outdated laws that have kept the U.S. energy economy creeping along at about 13 percent overall efficiency when it could double that without any new technology or new fossil-fuel supply. (Japan achieves about 20 percent, and we see a way to exceed that.) The effect would be to essentially double the country's energy service per unit of fuel burned. Sharply cutting the supply of fossil fuel needed would accelerate the hoped-for energy independence—and would greatly heighten energy security, which isn't always the same thing as independence. And by using less fuel to do more work, this strategy will sharply reduce carbon emissions.

The economic chasm we have to cross to reach that goal has two critical dimensions. First is its sheer breadth—the number of years it will take for wind, solar, and other clean renewables to replace the bulk of the fossil fuel supply we now depend on. Second is its depth—the depth of economic depression that must be overcome by a restoration of economic growth. Is there a strategy that will shorten the transition to renewables *and* stimulate growth? The energy-transition strategy we propose will help do both: It will bridge the chasm and shorten it.

As suggested by the Mittal energy-recycling case, more intelligent management of the existing fossil fuel supply can actually boost the productivity of the energy sector faster than the renewables bandwagon can. How? Increasing the *energy service* per unit of primary energy input proportionately reduces the cost of that service. And that reduction of cost, as we explain in this book, will drive economic growth.

The prevailing economic theory holds that growth is driven by capital investment and labor plus a very large, unquantifiable, factor—"technological progress," which remains "exogenous" (outside the forecasting calculations) because economists have been unable to fully identify or explain it. As a result, the capabilities of standard models to forecast economic growth have been notoriously poor. But new research has shown that the largest driver of growth is not so mysterious after all. The real engine of economic growth, it turns out, has been *the growing use (thanks to declining costs) of energy service, decade after decade.* (The term "energy service" refers to what economists with training in physical science call "useful work.") The proof, which we summarize briefly in this book and provide in full detail on our web site, is that incorporating the energy-as-useful-work factor into the economic models dramatically improves their long-term explanatory power—and forecasting capability.

The most exciting implication of this finding is not that the standard models need to be revised (they do); it's the more pragmatic prospect that a strategy that reduces the costs of energy services (by increasing output and profit per unit of fuel) will also help drive economic growth and recovery in the coming years. Our assessment suggests that we can engineer the "bridge" to achieve these goals by implementing eight proven (although in some cases little-publicized) technologies, of which the aforementioned waste-heat recycling is just one.

Two critical and potentially world-changing implications follow. First, the largely unquestioned assumption about how the cost of the multitrillion-dollar bailouts and recovery bills of 2008–2009 will be repaid—the assumption that the economy will soon recover its robust growth, as it always has in the past, because an infusion of new capital or spending power will drive it—may be wrong. Second, if low energy-service costs are needed to drive growth, the economic prospect is more dire than most experts have thought.[1] As global oil production peaks and begins to decline while the energy demands of China and other fast-growing countries continue to rise, and as climate-change constraints on all fossil fuels continue to tighten, fossil-fuel energy prices will rise higher than ever. Consequently, economic growth will halt or be thrown into reverse—*unless* we find ways to make energy services cheaper. If the nation's energy management in the coming years can double the productivity of its present supply, getting twice as much energy service or useful work (heat, light, propulsion, and so on) per barrel of supply as it does now, the cost of that service will decrease and growth can continue.

To explain more specifically how this can happen within the time span required for the transitional bridge, we take our assessment one important step further. Beyond challenging the currently prevailing theory of economic growth, which is critical to the recovery of the energy economy, our analysis suggests that there will be an unexpected silver lining to the darkening prospect of prolonged economic struggle being worsened by climatic destabilization. Experts on the ideological left and right might differ sharply in their projections of what climate mitigation will cost, but most agree that the cost will be significant and will reduce economic growth. However, our assessment suggests that we can accomplish significant parts of the transition strategy at a *negative* cost—simultaneously reducing energy costs, fuel use, and greenhouse emissions. Other parts of the bridge will involve low net costs that the nation can fund by shifting government support from currently unproductive projects to ones that are demonstrably productive.

This is where the strategy of building a bridge to the future using proven components becomes doubly important: It avoids huge capital costs (such as the construction costs of new nuclear or coal-burning central power plants, or oil-drilling platforms) that the United States

---

[1] Even if short-term energy prices fall, as they did in late 2008, thinking that the long-term threat has abated is a mistake. The price of a gallon of gas or a barrel of oil reflects current stockpiles, not global reserves, which will inexorably continue to shrink.

cannot afford to pay or wait for, and it cuts energy cost and drives economic growth by quickly increasing economic output per barrel of oil or oil-equivalent. We should note that oil man T. Boone Pickens did the country a service by publicizing the need for an energy bridge in his highly publicized campaign to increase subsidies for natural gas in 2008, but what he was asking for (more money to search for natural gas) would not have provided such a bridge.

This book is about what is needed to build that transitional bridge. Little or no ingenious new technology is needed, although development of the technologies required for that safer place on the far side of the bridge should of course continue to expand. Over the next few years, what's needed most is for those who see that pristine future so clearly to look down at the increasingly unstable economic ground right under our feet—and the economic chasm ahead—and to see just as clearly the outlines of the bridge we need to build before we can get to that safer place.

1

An American Awakening

Energy—that magical thing that enables vehicles to fly, Las Vegas to light up like a cosmic Christmas tree, or any basketball team or entertainer you choose to appear instantly in your living room—has always been abundant and cheap, at least for Americans. However, without *cheap* energy, our jet airplanes, electric lights, TV, and personal computers would be useless junk. When astute science-fiction writers envision a post-apocalyptic world, it isn't the technologies that have been lost, but the energy to run them.

Before Americans found oil in Pennsylvania, and later in Texas and California, and long before we relied on imports from the Middle East, we had abundant supplies of wood and coal. Wood is still the main source of cooking fuel for a third of the human population worldwide. But as land clearing and fire have deforested large swaths of the world, people who have no other major energy source have become desperate. Any country that loses its energy is at risk of social and economic breakdown. It can be the richest country in the world in invested capital and worker skills, but if its affordable energy disappears, it will likely see its way of life—and liberty—rapidly deteriorate.

Unlike North Africa, Ethiopia, Kurdistan, or Lebanon, the United States never lost all its vast forests, and made a fairly smooth transition from colonial times to the modern era of oil, gas, and nuclear power. In addition to endless forests, Americans were fortunate enough to have plentiful coal to drive steam locomotives, oil to run automobile engines, and rivers and waterfalls to produce electric power. For Americans, energy grew on trees—literally. So, right from the start, Americans took it for granted that the energy needed to drive our ingenious technologies would be either free or *cheap*. So unquestioned was this assumption that when the chairman of the Atomic Energy Commission, Lewis Strauss, famously promised in 1954 that "our children will enjoy in their homes electrical energy too cheap to meter," hardly anyone doubted him.

As a result of this deeply ingrained assumption, we have tended to monitor our economic progress in terms of the technologies we devise instead of the energy we produce to run them. "American ingenuity" is in our blood. Thomas Edison, Henry Ford, the Wright Brothers, and Bill Gates are our icons. And economists teach that "technological progress" is a major factor in economic growth, although they *have never been able to explain it,* except to call it an unexplained "residual."

Therein lies a mystery about the history of American economic policy that could soon come back to haunt us. To be more accurate, it's a mystery to the general public, but a puzzle to the theorists who have been the top economic advisors to American government and business for the past half-century. Behind the scenes, in their continuing efforts to turn economics into a true predictive science, economics professors have struggled with an uncomfortable fact: When they can't say exactly what "technological progress" really means or how it is produced, they're admitting to a large gap in their predictive capability—large enough to risk serious miscalculations in assessing how to cope with the coming disruptions of declining global oil production and accelerating climate change.

The gap was actually discovered in the early 1950s, when economists made their first quantitative reconstructions of historic economic growth during the previous century—since the Civil War. A key purpose of the reconstructions was to test the standard theory of economic growth to see how well it worked. According to the neoclassical theory of that time (which, in modified form, remains the dominant theory today), two factors of production drove economic growth: invested capital stock and labor supply. Endless squabbles have arisen about how to measure capital stock and even labor supply. But those two factors were, and still are, regarded as the drivers of growth.

The discovery came as something of a shock. According to the reconstructions, the accumulation of invested capital per worker accounted for only *one-seventh* of the economic growth that had actually occurred. That left roughly six-sevenths unaccounted for.

Robert Solow, the principal architect of the current theory of growth, who was awarded the Nobel Prize for his work, characterized the missing six-sevenths as "a measure of our ignorance." Others called it "the Solow residual." In academic discussions, it became referred to as total factor productivity (TFP), which made it sound as if the economists had identified something quite specific—but they hadn't. The most descriptive term was "technological progress," which was understood to mean the actions of innovative entrepreneurs who continuously create new ideas or inventions to keep spurring new economic activity. Because it couldn't be explained by economic variables, technological progress has been assumed to be **exogenous** (independent of economic forces). It's a short step from there to assume that past rates of growth will likely continue indefinitely into the future and that our grandchildren will consequently be much richer than we are

During the half-century after Solow's disconcerting discovery, the U.S. economy kept right on growing. So the fact that economists couldn't fully explain *why* it grew wasn't a big issue. However, today we have reason to think it's an issue of considerable importance, because it raises a question that we've never had to face: Can we safely assume that U.S. economic growth will continue for the next century at historical rates, even if cheap oil disappears from the menu? Is it safe to assume, as most economists still do, that "our grandchildren will be much richer than we are"?[1]

This is an important question because, if it's really true, maybe we Americans can soon resume borrowing from the increasing value of our homes and just enjoy our lives. If it's true, we can continue putting off necessary but expensive repairs to our environment. We can let the next generation do all that. Some conservative economists have even said that it's almost criminal to make such investments now because it would be like "the poor subsidizing the rich." But will the

---

[1] Behind the scenes, serious questions began to be raised in 1973–1975 when the Arab oil boycott triggered a major recession. If something was missing in explaining past growth, had the recession possibly exposed the missing factor as energy? Some economists tried including energy as a third factor of growth, but the results weren't sufficiently persuasive. In any case, oil prices came down in the 1980s and growth resumed.

next generation be rich if the cheap energy service runs out? Will they be rich if we don't save and invest? We think something is missing from the prevailing theory.

The problem is that the economic advisors who rely on this theory don't know what will drive economic growth during the period of the energy-transition bridge—a period that will be critical to the sustainability of twenty-first-century civilization. The advisors are heavily counting on some undefined third factor of economic growth to help get us back on track quickly. And because they can't define that third factor more precisely, the Obama administration has taken a shotgun approach: throwing new capital to bridges and highways, auto parts, "clean coal" research, carbon-capture experiments, housing, stem-cell research, space exploration, biotech, digital TV, and hundreds of other needy sectors, along with more tax cuts. They are nervously confident that *something* will pay off.

In 2009, White House Budget Director Peter Orszag told PBS interviewer Judy Woodruff, "I think we're anticipating the economy will be growing before 2011." Woodruff questioned whether President Obama's plan to increase taxes on high-income Americans, at a time when the economy was in crisis, might not impede instead of stimulate recovery—for example, by making it harder for business owners to "grow their companies and hire and create jobs." Orszag replied, "Well, again, I want to be clear about the timing here. This is in *2011*, and thereafter, after the economy presumably has begun to recover." Yet that presumption was (and is) a giant gamble, based on a very brave hope instead of any science-based economic theory Orszag could cite.

What *Does* Drive Economic Growth?

Around 1980, at the University of Wuerzburg, Germany, Reiner Kuemmel, a professor of theoretical physics, found himself pondering the economic consequences of the 1973–1974 Arab oil embargo and the 1979–1980 Iranian crisis. He found it curious that although energy was (to a physical scientist such as him) obviously critical to all economic activity, it was somehow missing from the prevailing theory of economic growth, as originally formulated by Robert Solow and subsequently adopted by virtually all economists. He strongly suspected that energy could be shown to be a factor of growth, along with labor and capital. Unlike others who had shared this suspicion but given up, Kuemmel made a serious effort to resolve it. With help from several young colleagues, he developed an alternative growth model that incorporated primary energy as a third explanatory variable. Not having been trained as an economist, he left out one of the assumptions that other economists routinely made—and still make. (We return to this later.)

In the same way that Solow had decades earlier tested his two-factor model to see how well it could reconstruct the growth that had occurred in the United States during the previous half-century, Kuemmel and his colleagues tested their three-factor model for three countries—the United States, the United Kingdom, and Germany—for the period from the end of World War II to 2000. Unlike Solow's model, which had never explained most of the growth that had actually occurred (and which Solow attributed to "technical progress"), Kuemmel's model appeared to mimic the actual history almost perfectly. However, the parameters of his model were hard for economists to interpret, and most economists were skeptical. One remarked that, with enough parameters, a mathematical function could reproduce *any* shape, even that of an elephant.

Kuemmel was not alone. Working independently, the senior author of this book (Robert Ayres, a professor of environment and management at the European business school INSEAD in Fontainebleau) was following a different track but with the same basic intuition. Ayres was also trained in physics and had spent a large part of his career studying energy and material flows, technological change, and input-output economic models from an environmental perspective. With his research assistant, Benjamin Warr, he decided to construct a three-factor model that focused not on primary energy, per se, but on the thermodynamic efficiency with which primary energy (**exergy**) is converted to "useful work."

Ayres and Warr then undertook the laborious task of reconstructing the historical data on useful work output for each country, starting with the United States. They tested the new model formulation by reconstructing economic growth during the entire twentieth century, initially for the United States, and later for Japan and the United Kingdom, and for Austria since 1920. The results were dramatic: The new approach seemed to explain nearly 100 percent of twentieth-century

economic growth for each of the four countries. Again, the few economists who saw the results were skeptical. The results were "too good to
be true." Both the Kuemmel model and the Ayres–Warr model had
ignored a certain article of faith among trained economists: that the relative importance (known as **output elasticities**) of capital and labor as
factors of growth should be exactly proportional to their "cost shares" in
the national accounts.

The cost-share assumption came from a simple model calculation
that had gone unchallenged for decades. However, unlike the abstract,
oversimplified economy implied by the model, the real-world economy
is not a single sector producing a single product. And Kuemmel has
recently provided mathematical proof that the output elasticities don't
need to equal cost shares in a realistic multisector economy. His proof
answers the primary objection of the mainstream economists. Readers
can find this proof, along with a statistical analysis that responds to the
objections that the reconstructions provided by the two independent
models are "too good to be true," in greater detail in several publications and on our web site.[2]

The Ayres–Warr growth model seems to clear up some of the long-
standing uncertainty around "technological progress"—the missing
third factor of growth—by identifying most of it very specifically as *the
increasing thermodynamic efficiency with which energy and raw materials are converted into useful work.* If economists can accelerate this
efficiency by applying smart policies, economic growth will also accelerate. Conversely, if the rate of efficiency gains slows in the future, economic growth is almost certain to slow as well. The fallout of this
discovery, combined with the eye-opening results of the new growth
models, provides what we believe is a significant new perspective on
the nature of economic productivity and growth. It also sheds light on
how that growth (or lack thereof, if the cost of energy services continues to rise unchecked) will affect our efforts to make a successful
energy transition during the coming decades.

---

[2] See Robert U. Ayres and Benjamin Warr, *The Economic Growth Engine: How
Energy and Work Drive Material Prosperity* (Cheltenham, U.K., and
Northampton, MA: Edward Elgar Publishing, 2009). Also see Reiner Kuemmel,
Robert U. Ayres, and Dietmar Lindenberger, *Technological Constraints and
Shadow Prices* (Wuerzburg, Germany: 2008), on our web site.

The story of how an economic paradigm could be so incomplete that it cannot account for a primary driver of economic vitality and growth is rooted in the eighteenth and early nineteenth centuries, when natural resources were lumped under the category of "land." Landowners managed the economy (and landowners had the vote). Later in the century, land was absorbed into the larger category of "capital." In an agricultural economy, economic productivity (gross domestic product [GDP]) was mainly the output of farmers working the land—laborers working with capital. And in the mechanized economy of the Industrial Revolution, economic productivity constituted factory workers running machines.

What about energy? The energy input to an agricultural economy was the sunlight needed to grow plants (which, in turn, provided feed for the working animals). But the amount of sunlight was proportional to the acreage of the land, so a natural part of owning land was having the sunlight that went with it. The theory didn't need to count energy as something separate from capital. The two-factor (labor and capital) theory became central to the thinking and teaching of economists in the nineteenth century and has remained so until now.

Perhaps the first energy crisis of the 1970s should have been a wakeup call, a warning that the theory needed to be revised, but it wasn't. Primary energy was—and still is—treated as an intermediate *product* of labor and capital. A combination of labor and capital is presumed to somehow produce energy. Coal miners go to work lopping the top off a mountain in West Virginia and "produce" coal. That might seem to make sense if you think of a coal mine (or an oil well, or a wind farm) as a producer of new energy. But it's not. The useful energy is already there—in the mountain, the sea bottom, or the wind. Maybe that's why physical scientists, for whom the first law of thermodynamics is not so easy to overlook, began to question the two-factor theory. Labor and capital *extract* energy; they don't *make* it.

The difference between extracting energy and producing it might sound like a semantic quibble. But there's a huge difference between energy being a *product* of economic activity, as the labor-and-capital theory claims, and energy being a *prerequisite* to economic activity. First, economic activity can't happen unless both the labor and the capital are fueled. Laborers need to eat, and capital needs to be fed.

If the capital is a farm field, it needs that sunlight; if it's a machine, it needs fuel or electric power. Without the prerequisite energy, neither the labor nor the capital can produce one dollar of GDP. This distinction between energy as an intermediate product in the two-factor theory and energy as an independent third factor is important because *the resulting calculations of economic productivity and growth are radically different.*

In the economic crash of 2008–2009, we witnessed an avalanche of signs that the edifice of the dominant economic paradigm had cracked. However, the edifice wasn't just an obsolescent artifact of the eighteenth-century origins of capital as tracts of land with their energizing sunlight already given. It had also grown out of a general inclination, among managers of business and government alike, to take the availability of physical resources largely for granted. Investing was like buying toys with batteries included. To a physical scientist, however, the economic story needs to begin with energy resources. The story of technological progress, as taught in our schools, might understandably focus on ingenious inventions, and on the mind-boggling progressions we have experienced, from steam power to electric power, carriages to cars, and typewriters to computers. The latest iPod is intrinsically more interesting than a lump of coal, although a lump of coal still energizes it. As we stressed earlier in this chapter, without fuel to feed them, the machines are useless junk.

The history of machines, then, is also very much the history of the fuels that feed them. The fact that both the sources of energy and the machines themselves have become almost continuously *cheaper* throughout the past two centuries—until very recently—has reinforced the substitution of fossil (and nuclear) energy for human and animal muscles.[3] But as fossil energy becomes more expensive in the post-peak-oil, carbon-constrained future, the energy-intensive sectors, including chemicals, metals, transportation, and all manufacturing

---

[3] Crude oil prices, in 2006 dollars, dropped from more than $80 a barrel in 1869 to less than $20 a century later in 1970. The OPEC oil embargo of 1973 temporarily interrupted the downward trend, but the trend resumed in the 1980s and 1990s. However, the trend was interrupted again after 9/11, and most analysts—including us—see no prospect of oil ever again being cheap over the long term.

and construction, will slow down. This will happen no matter what ingenious new products or inventions are brought to market—*unless new technologies also make energy services cheaper.* And if energy services delivered to users get *very* expensive, economic growth will stop altogether, with probably disastrous social consequences.

The complicated mathematical models economic advisors use to guide economic policy, with their traditional focus on labor and capital as drivers, pay too little attention to the details of how primary energy puts labor and capital to work. In the early Industrial Revolution, when steam engines started replacing human and animal muscle power, the evolving economic theory never fully recognized and incorporated some basic facts of both animal and industrial metabolism:

- Horses, oxen, and other working animals *eat plants* for Calories.[4] Moreover, while they are working, they can't feed themselves by grazing; they need to be fed harvested grain.
- Human laborers can't graze, either. They eat food products from plants and the flesh of animals that eat plants, also for Calories.
- Machines *"eat" fossil plants* (coal, oil, and natural gas derived from plants that grew in the Carboniferous Age), also for Calories.

Consequently, today's economic advisors have too easily ignored a basic fact of civilization that might be critical to the world's growing troubles: *All economic activity begins with physical materials and energy carriers (fuels and electric power).* Without materials, there can be no food, shelter, or technology; without energy, there is no work—and no economic activity.

Why does this matter? If the economic models are revised to properly account for the critical importance of materials and energy in economic forecasting, the results will show a radically different future for the American economy than the standard models presently do. Matters of theory might seem of little immediate interest to people who have lost their houses (whether seized by the bank or

---

[4] Food Calories are usually referred to with a capital *C*, to distinguish them from heat calories. One Calorie is equal to 1,000 calories, or 1 kilocalorie.

destroyed by flood or tornado) or their money (whether from a layoff or a hedge-fund collapse). But without a strong theoretical understanding of how the world works, we'd still be living in caves and chasing rabbits for food. Uneducated people couldn't have invented cars, airplanes, or computers, nor could these inventions have occurred by accident; they had to be preceded by discoveries of the underlying principles of physics, chemistry, metallurgy, and mathematics. If the theory guiding economic policy has been missing a key part, sooner or later the engine of growth in the real world is likely to crash. We may have reached that point.

An Economic Perfect Storm

When the housing market began to collapse in 2007, economists' first response was to reassure the public that there was no need for panic because "the economy is sound." But that reassurance was more a reflection of the economists' faith in free-market capitalist ideology than of their ability to predict what would really happen. If we look instead to the forecasts of physical and resource science, the outlook is very different. We are in the early stages of a kind of global "perfect storm" that will affect everything we do, and for which we are woefully unprepared. How well we design the energy-transition bridge will profoundly affect how well our country—and civilization itself—endures in the twenty-first century.

The storm isn't invisible, but its coming impacts may still lie beyond the consciousness of most Americans—because we've been subjected to massive campaigns to keep these impacts out of sight and out of mind, and because of pervasive myths about the nature of energy economics that have kept government and business leaders persistently looking in the wrong direction.

The symptoms of the storm are both economic and physical, and the connections between these two sets of signs are critical. The economic signs of destabilization include not only the erratic spikes we have seen in the prices of gasoline and crude oil, but also—if and when growth resumes—the rising long-term costs of everything dependent on oil and gas, from food to plastics to air travel.[5] The physical signs include the escalation of extreme weather disasters that we have experienced in the past few years, both worldwide and in the United States. The Myanmar cyclone of 2008 killed more than 85,000 people, nearly 40 times the number who died in the 9/11 Al Qaeda attack on the United States. That particular storm may or may not have been a consequence of global climate change, but the increasing frequency and ferocity of such storms almost certainly is. In a single day in 2008, the United States was struck by 87 tornadoes. In a single month in 2008, 2,000 wildfires raced out of control in California, incinerating more than 1,400 square miles.

These disturbing events, along with similar events on other continents, are the harbingers of three imminent and irreversible phenomena. First is the aftermath of "peak oil," that fateful moment when the public is finally convinced of what experts will have known for some time: that global oil production has begun its final decline. The global economy, built around oil, will then have entered a time of permanently worsening shortages and spiking prices—even as the global population, with its fast-rising demand for energy and food, continues to expand by more than 70 million people per year. Most forecasts indicate that peak oil will have occured sometime between 2010 and 2020, and some experts think it has happened already.

The second disruptive phenomenon, following a slowing of energy-related innovation during the past half-century, will be the conspicuous "old age" of several key technologies on which the fossil fuel–based economy has depended. Neither the internal combustion engines produced today nor the steam turbines we use to generate our electric power are much more efficient (at converting fuels to useful energy) than they were in the 1960s. Our "rust belt" is rusting for a *reason*.

The third disruption—which most Americans now see coming but don't quite know what to do about—is the escalation of climate catastrophes, which will compound the U.S. economic and energy

---

[5] Temporary slumps in demand, such as those caused by the economic collapse that began in 2008, bring temporary reductions in energy prices, but the long-term trend—given the magnitudes of expected global population growth coinciding with declining oil production—will be rising costs of oil.

crisis by increasing the urgency both to achieve oil independence and to fully develop our new, non–fossil fuel technologies—which, contrary to some impressions, are nowhere near ready to take over on a large scale. Ironically, the emergency costs of coping with climate catastrophes in the short term could divert money from investment in the very technologies needed to avert still greater catastrophes during the longer term.

It's not hard to see what the bottom-line response of the United States to this storm must be. We can summarize it in terms of three main objectives. In later chapters, we show how, if the economic model is corrected so that it fully accounts for the impacts of energy flows on economic growth, we can—in relatively few years—achieve each of these objectives:

- **Massive reduction of greenhouse gas emissions** *at negative or very low net cost.* Negative-cost options will actually boost the American economy as a whole.

- **Energy independence** without drilling for oil off the California or Florida coasts or in the Arctic National Wildlife Refuge, and without the need for military control of Iraq, Saudi Arabia, Venezuela, or any other place that might tempt our leaders if the price of oil gets too high, too fast.

- **Greatly improved energy security** (which is not the same thing as energy independence), if central power plants and transmission lines are vulnerable to sabotage or tornadoes, as they are today.

Paralysis of Leadership

Americans who believe that free markets will take care of everything might be in a state of gloom or denial right now, but others might be hoping that our business and government leaders are finally "getting it." With all the news of new green initiatives—hybrid cars, electric cars, biofuels, compact fluorescent lights, Energy Star appliances, voluntary carbon markets, and green investment, along with the Obama administration's greatly augmented investments in renewable energy—it's easy to conclude that America is responding vigorously. And, yes, many individual citizens, companies, and communities are working hard to reduce carbon footprints and prepare for climate change. Yet the U.S. government under the Clinton and Bush administrations did virtually nothing, and, contrary to some misleading advertising, most of the largest corporations have done little apart from advertising. *Carbon dioxide emissions continue to rise.* Scientists who monitor the situation clearly see that, even with the intervention of the American Recovery and Investment Act of 2009 and a proliferation of clean-energy investments, the carbon dioxide concentration in the atmosphere will *continue* to rise for years to come, unless more fundamental changes are made in our energy economy.

The second Bush administration awakened very late to the approach of the three phenomena that constitute the coming perfect storm. For seven of his eight years as president, George W. Bush and his closest advisors denied that human-caused global warming was real and took vigorous action to *stop* efforts to prepare for the impacts. But after the devastating Midwestern tornadoes and floods of 2008, the White House issued a statement acknowledging what could no longer be denied: Climate change is at least partly driven by human activities.

Meanwhile, however, Oklahoma Senator James Inhofe and some other oil-country politicians have continued to insist that global warming is a "liberal hoax." Their vehement protests—combined with considerable confusion among legislators about whom to listen to (climate scientists or Exxon and Shell lobbyists?)—were enough to keep Congress from passing any meaningful climate-related legislation before 2009. Political adversaries of former vice president Al Gore had long ridiculed and pilloried him for his assertion that the internal-combustion automobile engine should be phased out. By the time his film *An Inconvenient Truth* helped to raise public awareness of the problem, nearly two decades had been lost.[6]

The expected approach of peak oil has been widely discussed behind the scenes, mainly by some oil geologists—notably King Hubbert, Colin Campbell, and Jean LaHerriere. But most energy

---

[6] NASA's chief climate scientist, James Hansen, first warned the U.S. Congress about the dangers of global warming in 1988. He reiterated the warning in 2008, but with the comment, "Now, as then, I can assert that these conclusions have a certainty exceeding 99 percent. The difference is that now we have used up all the slack in the schedule."

economists, the International Energy Agency, OPEC, and the U.S. Energy Information Administration, along with the oil and gas industry associations and the big oil companies themselves, have long promoted the notion that demand for oil could and would continue to rise without significant price increases until at least 2030. Although recent predictions are more cautious, some economists still say "there is an ocean of oil; it is only a question of price," citing tar sands and other questionable sources as justification. The net effect of all this institutional optimism in the face of geological skeptics has been legislative and administrative paralysis.

Scientists began warning—as long ago as the 1850s, and more seriously in the 1950s—that the planet's resource base is not unlimited. Apart from worries about the availability of coal and oil and other mineral resources, agronomists have been concerned that soil erosion is taking away the topsoil needed for human food production faster than nature can regenerate it. Biologists such as Edward O. Wilson have warned that pollution and habitat destruction are threatening tens of thousands of species—including honeybees, which are essential to the pollination of about half of all human food crops. Ecologists have observed that as species are decimated, entire ecosystems are breaking down. Marine biologists have noted that sewage from coastal cities and fertilizer runoff from farms have turned large areas of the Gulf of Mexico and Atlantic Ocean into anaerobic dead zones. In the 1970s, climate scientists began to worry that carbon dioxide and other so-called greenhouse gases (GHGs) rising from human activity might destabilize the climate.

In the early 1990s, the worries coalesced into explicit warnings:

- In 1992, the climate scientists of the newly formed Intergovernmental Panel on Climate Change (IPCC) issued their First Assessment Report, warning the governments of the world that human-exacerbated climate change was becoming a serious threat to civilization. The report was prepared by 78 lead authors and 400 contributing authors from 26 countries, reviewed by 500 additional scientists from 40 countries, and re-reviewed by 177 delegates from every national academy of science on Earth.

- In the same year, a broader assemblage of 1,670 scientists from a wide range of fields issued a report called the *World Scientists' Warning to Humanity*, noting in its Introduction that

"Human beings and the natural world are on a collision course." The document's first recommendation was, "We must...move away from fossil fuels to more benign, inexhaustible energy resources to cut greenhouse gas emissions." The warning was signed by 104 Nobel Prize winners.

- In 1998, the International Union for the Conservation of Nature (IUCN) issued its 862-page *IUCN Red List of Threatened Plants,* summarizing 20 years of research by 16 science organizations worldwide. The *Red List* showed that 34,000 of the world's known species of plants were approaching extinction. A national survey of American biologists by the American Museum of Natural History in New York concluded that we have entered the fastest mass extinction in Earth's history—even faster than when the dinosaurs died 65 million years ago.

- In 2007, the IPCC issued an updated assessment, revising its warnings about climate change based on more recent and ongoing studies. The new report showed that temperatures will rise higher, sea levels will rise higher, and catastrophic damage will probably be more severe than had been indicated in its earlier assessments.

Many economists have had a very different take on the situation. While not disputing that climate change might be approaching, they have seemed far more concerned about the near-term costs of buying insurance against the approaching changes, than with uncertain costs of damages done in the future. Perhaps that is partly because economists are trained to discount the future, largely based on the argument that individuals and businesses demonstrably do so. After all, society is essentially a collection of individuals and businesses. Discounting is partly due to mortality—we might not live long enough to enjoy what we have saved from current consumption—and partly due to inherent shortsightedness. We find it hard to imagine that, as individuals or as society, we might have greater needs in the future than we do today. If economic growth is guaranteed, why not let the richer folks of the future pay for cleaning up the environment because they will be better able to afford it?

How much to discount[7] the needs of the future is part of an immediate and politically pressing question: How much should we invest today to ameliorate climate damages that will happen 5, 10, or 50 years from now? The answer surely depends, to a significant degree, on the future state of the economy. If climate change will make the world richer than it otherwise would be—by reducing the need for heating oil, and enabling longer growing seasons in places such as Siberia— then we might even want to accelerate that change. Conversely, if climate change will make the world poorer because of rising sea levels, more violent storms, more hurricanes and tornados, more devastating floods and droughts, and more widespread crop failures (and science emphatically points to that likelihood), we should invest in strategies to reduce carbon emissions and mitigate those effects. To decide what to do, *we need to know how changes wrought by peak oil and climate change will affect the economic system.*

But the standard economic models cannot begin to answer such questions without cheating. As we have noted, they cannot explain most of the economic growth that has occurred during the past two centuries. What if that growth has been mainly due to declining costs of primary energy and useful work? What if those costs will likely rise in the future? Labeling the main cause of that growth "technological progress" or "total factor productivity" evades the question. It doesn't explain how much growth—if any—we can expect during the next 5, 10, or 50 years. As we have mentioned, the standard models of economic growth simply assume that growth will continue at the same basic rate, or slightly less, unaffected by what we now know to be the critical role of energy availability and cost. The models also assume that all business units and industries function on a level playing field: They enjoy perfect weather, perfect adherence to the rules, honest

---

[7] There is much dispute among Economists about what the social discount rate should be. Some heterodox economists believe it should really be zero, or a very small number, because future generations should have a vote in our current decision making, since they are the ones who will have to live with the consequences. But most economists tend to think in financial terms, arguing that a dollar earned can be spent today or saved in the bank, where it will earn interest (presumably greater than the rate of inflation) and, therefore, have greater buying power in the future.

umpires, an orderly and well-behaved crowd in the stands, and hot dogs for all. In stark contrast, the warnings of physical science describe a coming time in which extreme weather events will combine with weakened ecosystems, a growing human population, poverty, resource depletion, and increasingly frantic struggles over water and energy, to produce enormous human disasters that wreak physical, social, and economic havoc. The standard economic model has no way of accounting for havoc.

The economists (but until recently not the scientists) have heavily influenced the U.S. government, partly because economic advisors in recent years have been like the courtiers in the tale of *The Emperor's New Clothes*. Although they have known since the 1950s that the "technological progress" factor in economic growth was unexplained, the serendipitous growth of the economy—and its repeated recoveries from setbacks—gave them cover. And as a profession, they kept quiet about what they didn't know. Like the naked emperor who thought he was magnificently dressed, the President of the United States and his corporate friends were assured by their advisors, at least until 2008, that all was well. One of the most frequently offered assessments, especially during recessions or slumps (from which the U.S. economy has always seemed to rebound vigorously) was—as we kept hearing during the collapse of 2008—that "the economy is fundamentally sound" or "the fundamentals are sound."

But as most people now know, and as we elucidate in this book, the economy was *not* sound: We were living much too high off the hog, in several ways. We were—and are—borrowing and spending natural capital as if it were current income. Much of that natural capital is irreplaceable. During the past few decades, we have also been spending an immense amount of borrowed money for current consumption that the next generation of Americans won't likely be able to repay. The notion frequently promulgated by conservative economists, that our grandchildren will be much richer than we are, looks increasingly like the emperor's nonexistent clothes.

We can see a glimpse of the dynamics of the economic confusion that now reigns in the public accusations and arguments that took place in the wake of the 2008 housing crisis. Few experts had seen it coming, and virtually none were in high positions. Even after the crisis started, few anticipated its ripple (and then tidal-wave) effects. When the subprime crisis first hit, Federal Reserve Chairman Ben Bernanke reassured the country, "Given the fundamental factors in place...we do not expect significant spillover from the subprime sector to the rest of the economy or to the financial system." Within a few months, the crisis had spilled over to the whole mortgage industry, home construction, real estate, insurance, investment banks, and auto companies. When news broke in July 2008 that a major California bank had failed and that U.S. mortgage giants Fannie Mae and Freddie Mac were near collapse, a White House spokesman went on TV and said, "There's no reason to panic. Fannie Mae and Freddie Mac have plenty of capital."[8]

No one we know of pointed to that comment as indicative of anything but the Bush administration's hope of assuring the nervous public that the economy was sound. ("The economy is sound," President Bush announced a few days later, in another of his "mission accomplished" declarations.) But in fact, that reassurance about Freddie and Fannie having "plenty of capital" offers an unintended demonstration of the economic confusion and misdirection that has stymied U.S. energy policy. The statement was intended to mean that the lenders had plenty of the essential asset needed to underpin credit and restore economic vitality—*capital*. As we noted earlier in this chapter, the neoclassical theory always presumed (at least until the 1950s) that it is invested capital that fuels the economy. But as we also noted, it is *energy* that fuels the economy. Maybe there was indeed enough capital to keep the housing market solvent—after another capital infusion from the taxpayers. But was that enough to put the housing market back on its feet?

The answer depends on the state of the economy at large. Demand for new housing depends on demographics, but it also depends on people wanting to move into a bigger or better house, in a more desirable location—and that depends on economic growth. Contrary to what we're told by the standard economic model, economic growth depends on the availability of ever-cheaper energy services or useful work.

---

[8] According to Federal Deposit Insurance Corporation (FDIC) records, 25 banks had failed in the United States by the end of 2008. The failures continued at an even faster rate in 2009.

The U.S. government can't print energy the way it can print money. It can't do accounting tricks to make energy appear, as it can make capital appear. The reality is that the U.S. economy is *not* sound, because energy is no longer cheap. And although the price of gas or oil might drop temporarily as it did in late 2008 (because of temporarily reduced demand), no plausible long-term scenario will ever make primary energy cheap again. However, the United States *can* adopt a strategy that makes *energy services* cheaper than they are now.

Dispelling Energy Myths

The last years of the second Bush administration were marked by an air of discomfiture and hollow bravado in the government, and growing anxiety among the public. The anxiety arose not only from the continuing conflict between physical science and economics, but also from certain deeply ingrained myths and misconceptions about the U.S. energy economy, including the following:

- *Only by expanding oil drilling to the coastal waters of California or Florida, or to the Arctic National Wildlife Refuge, or by building hundreds of new coal-burning or nuclear power plants, can the United States achieve energy independence.* Objective analysis from numerous studies makes it clear that this is political rhetoric with no factual basis. The United States has been consuming more oil than it has produced domestically since 1970, and, as many experts have testified, the proposed offshore drilling cannot change that. The path to energy independence lies in the institutional and legal structure of the U.S. energy system, not under the ocean bottom.

- *Central power plants are optimally efficient.* In reality, our power plants throw away—as waste heat—twice as much of the energy that goes into them as they deliver to American consumers in the form of electricity. Yet if a key change were made in the institutional structure of the electric power industry, that waste heat could be used to eliminate the bulk of the fossil-fuel consumption now used to heat American buildings and homes.

- *Taking serious action to achieve greater energy efficiency and reduce carbon emissions to mitigate climate change will be too costly to businesses and will "harm the U.S. economy."* We provide abundant real-world evidence that U.S. businesses can achieve substantially reduced carbon emissions at negative costs—with rapid payback and, for tens of thousands of American companies, with the promise of continued profits for the next two decades or longer. This is possible without any need for exotic or expensive new technologies on the production side. The importance of that two-decade-or-longer bridge is that it's long enough to safely guide our economy through the very treacherous disruptions that we face between now and the time when the energy technologies of the future are ready to roll.

- *The main drivers of economic growth are labor and capital.* As we mentioned earlier and explain further in the next two chapters the largest driver of growth in the U.S. economy in recent decades has been the declining costs of energy services (useful work). With the price of oil (and gas) going up, a new energy strategy is essential.

A Science-Based Energy Strategy

In recent years, predictions of a great awakening—a phenomenon that profoundly transforms humanity—have proliferated. Hundreds of writers and thinkers have predicted that world-transforming changes of consciousness will save us from catastrophe. Some of this prophesying might stem from a yearning to believe that there's more to life than the dominant consumer culture seems to offer. Some envision an evolutionary leap from our species' historic preoccupations with war and ideological conflict to a higher state. Much stems from the environmental movement of the past few decades, and its insights about the interdependence of all life and the unsustainability of the present course.

Our own perspective is more mundane: We foresee a time in the next few years when Americans awaken to the realization that past misconceptions about the role of energy in our lives have been blocking us from preparing for the perfect storm that is rising around us. When we become fully conscious of that, we can achieve the following results:

- The U.S. response to climate change, both governmental and private sector, can shift from its recent crippling indecisiveness to a World War II–scale mobilization. An intelligently prioritized near-term investment in carbon emissions reductions across the economy can result in billions of dollars of expenditures preventing trillions of dollars of damage in the coming decades.

- Economists can revise the standard economic model used to advise government and business, to fully account for the role of energy in driving economic growth, and to convince policymakers that the highest priority for short-term government support must be to develop *the specific kinds of technologies that make energy services cheaper.* Other large categories of heavy spending, such as the military budget and public entitlements, might need to shrink until lowered energy costs can regenerate vigorous economic growth.

We can achieve these results even as we approach—and pass through—the perfect storm of social and economic disruption in the aftermath of peak oil, the old age and decline of key oil-based technologies, and the rising toll of extreme-weather disasters. Climate and energy policy will rise to the top of the nation's political and planning agendas. It will surpass even terrorism, "the next war," or the economy *per se* on the priorities list because it will finally be clear to us that cheap and secure energy services are the root of the economy and its capacity to attend to all other needs.

2

Recapturing Lost Energy

Politically and emotionally, energy independence has become a hot issue not only for Americans, but for oil-dependent countries all over the world. In 1973, the Arab oil embargo caused long lines at American gas pumps. In winter 2009, eight European countries had to go weeks without natural gas—causing millions of people to freeze—because Russian politicians decided to cut off their supply. Only a few countries are oil or gas exporters; the rest (including the United States) are increasingly at the mercy of those few—*unless* they can find a way out.

American politicians' responses to the call for energy independence have been reflexively quick and predictably consistent with their ideological proclivities. With the gasoline price spike of 2008, Republicans aggressively renewed their call to drill for more oil off the California and Florida coasts and in the Arctic National Wildlife Refuge, areas where drilling has been prohibited for environmental reasons. They also called for reviving the nuclear industry and building many new nuclear power plants. Drilling would be consistent with the long-held conservative view that the exploration and conquest of nature has been at the heart of the American quest,[1] and that the government shouldn't tell corporations what they can or can't do. Conservatives also argue that nuclear power wouldn't generate greenhouse gases. Environmentalists and the Obama administration have called for a shift from oil to renewable energy resources as fast as possible, because of the damage done in recent years both to the climate and to the nation's reputation (and clout) around the world.

---

[1] TV commercials by Exxon-Mobil in summer 2008 intoned that, in the great quest of the human experience, "we are explorers" by nature, and suggested that oil exploration on the ocean bottom is an inexorable extension of that noble quest.

Unfortunately, both of these political impulses are mistaken. The conservative call to drill for more oil in ecologically vulnerable areas is misconceived for two reasons. First, geological studies have made it clear that little oil would likely be found there[2]—the call is largely symbolic. And whatever is there will take a decade to extract, so the immediate benefit would be essentially zero. Second, it's possible to achieve U.S. energy independence *without* such drilling—and without the commensurate increases in global-warming carbon dioxide emissions that the extra oil produced would then generate. As we show in this chapter, we can make the fossil fuels that we're currently using produce more energy service—so much that, within the next 20 years, it will be possible to end oil imports from the Middle East without any new drilling off Palm Beach or La Jolla, or in the middle of a caribou migration route. The heavy lobbying for nuclear power tends to obscure the fact that, although nuclear sources provide some electric power, they don't provide a substitute for petroleum, either gasoline or petrochemicals.

Some of the environmentalists are mistaken, too. Although the need to replace oil and coal (and possibly nuclear power) could hardly be more critical, it will take at least several decades to make the full changeover. We share the goals—and the sense of urgency—of the alternative-energy advocates. But there is no politically or financially viable way to overcome the real-world constraints of capital depreciation, massive capital replacement of obsolescent fossil-fuel infrastructure (including roads and highways), and the impossibility of mobilizing new investment overnight. A gargantuan share of U.S. assets is locked into the old system; even under emergency conditions, it will take many years to free them up. On the other hand, even if the old infrastructure could be dismantled in a week, it would be a huge mistake because, paradoxically, the fastest way to achieve U.S. energy independence and sharply cut carbon emissions is to leave the old system in place a while longer—investing in short-term modifications that can greatly increase the total output of useful work with existing fuel inputs *and* simultaneously reduce the output of

---

[2] According to the U.S. Energy Department, the United States (including its coastal areas) has less than 3 percent of the world's known petroleum reserves.

greenhouse gas emissions. We can explain this best by looking at a real-world case.

The Hidden Gold of Energy Recycling

On the south shore of Lake Michigan, in the northwest corner of Indiana, the Mittal Steel Company has a coking facility called Cokenergy. Coke (the industrial substance, not the soft drink) is nearly pure carbon, made by heating coal in the absence of air to remove the methane, sulfur, ammonia, tar, and other impurities to make it suitable for use in a steel-making blast furnace. Some of the gas removed in this process is used to heat the ovens. In a conventional facility, the combustible coke-oven gas is captured, but the hot combustion products from heating the ovens themselves are normally blown into the air.

But Cokenergy is not conventional. In addition to recovering the gases for use elsewhere, this plant captures waste heat and uses it to generate electricity as a byproduct. This "recycled" energy is produced *without any incremental carbon dioxide emissions or other pollution.* Although the primary process (making coke) uses a fossil fuel, the subsequent production of electric power from the high-temperature waste heat does not. The byproduct electricity is as clean as if it were made by solar collectors. This carbon-free electricity is then used to run the rolling machines in Mittal's adjacent steel plant.

In 2005, the Mittal coking plant generated 90 megawatts (MW) of emissions-free electric power. As we noted in the Introduction, that output, combined with the 100MW of recycled energy that nearby rival U.S. Steel produced, exceeded the entire U.S. output of solar-photovoltaic (PV) energy that year. Combined with the more than 900MW of recycled waste-energy streams other American plants harnessed, the nation's recycled-energy output was about *seven times* the U.S. solar-photovoltaic production that year. Moreover, the companies that recycle their waste energy haven't needed to buy this power from local utilities. This eliminates all the carbon dioxide emissions (and other pollutants) that the utilities' production of that amount of power would otherwise generate. Yet the total U.S. production of emissions-free "bonus" electricity by this method is still only about 10 percent of the amount that currently operating American plants *could* produce—without burning any additional fossil fuel.

Solar PV has gained rapidly since 2005, but even if it continues to expand at a meteoric rate, it has started from such a small base that it will take many years to replace a large share of the fossil fuel we now depend on. Wind power is further along, but it, too, will need many years. And it's those "bridge" years we need to be concerned about. Companies can install facilities such as the one at Mittal Steel's Cokenergy plant within three or four years. And those facilities are profitable. The electricity from Mittal's recycling operation costs only half of what the local utility charges its customers.

It's a bizarre, perhaps ironic situation, to be sure. From an aesthetic or emotional standpoint, a progressive environmentalist might find it hard to accept that *using fossil fuel more effectively* is preferable to just switching as soon as possible to renewables, as so many people seem to suggest. But from the standpoint of physical science and engineering, it's indisputable: If our goal is to reduce carbon emissions on a large scale as quickly as possible, the most effective way is to invest in "cogeneration." This means recycling the high-temperature waste heat energy not just from coking, but from a spectrum of existing fossil fuel–burning industrial processes—such as smelting, oil refining, carbon-black production, and chemical processing—into electricity that's as clean as if it had been produced by wind or the sun. And this energy is cheaper.

That last point is critical: Recycling waste-energy streams from industrial uses of fossil fuels is still far *cheaper* than energy from solar-photovoltaic generation or wind turbines, and far cleaner than energy from biomass. The day will come when the renewables will be competitive without subsidies, and civilization will be on safer ground. Wind power is sufficiently developed to compete with nuclear power or fossil fuels in some windy places, but solar power (both thermal and photovoltaic) still has a long way to go. For the next few years, even with the 2009 financial rescue plan's boost for alternative energy, a dollar invested in waste-energy recycling such as the program at the Mittal plant will produce more emissions-free new power—and carbon dioxide reduction—than a dollar invested in renewables.

We must quickly add that this does *not* mean investors should have second thoughts about investing in renewable energy. For the strategy outlined here to make any sense, investment in solar, wind, and hydrogen sources should continue to increase. Energy recycling such as the kind Mittal Steel is doing is a short-term strategy intended to hold the fort until renewable output is big enough to take over. Until then, recycling the heat from the coke plant is the smartest thing Mittal Steel can do.

Unfortunately, this doesn't mean that such low-cost, emissions-free energy can provide the power for your home or office—yet. Mittal Steel distributes its 90MW from Cokenergy only to its own steel plant, not to the people of East Chicago, Indiana, where the plant is located. However, supplying clean electricity to the enormously energy-consuming steel-making process in this way not only reduces the need for Mittal to buy electricity from its local utility, but also greatly reduces the amount of carbon dioxide that the utility pumps into the air over northern Indiana.

In addition to high-temperature heat, we can recycle several other kinds of waste-energy streams that thousands of American industrial plants generate. We can inexpensively convert much of this waste to electric power that would otherwise need to be generated by coal- or natural gas–burning power plants or by nuclear plants.[3]

In Rochester, New York, the Kodak Corporation has a complex that stretches 5 miles end-to-end. A steam-pressure system that powers its chemical processing now recycles 3 million pounds of what would otherwise be waste *steam* per hour, generating electric power that, at last count, was eliminating 3.6 million barrels of oil-equivalent per year and saving Kodak $80 million on its electric bill.

A third category of waste-energy stream is flammable *gas*, which petroleum refineries and some chemical plants often simply burn off (flare) into the sky. If you've ever driven along a certain stretch of the New Jersey Turnpike at night, along I-95 near Philadelphia, or in the "Cancer Alley" area of Louisiana, you've seen (and smelled) a lot of gas flaring. In principle, companies could have used all that wasted energy to make cheap electricity.

---

[3] Nuclear plants don't emit carbon dioxide, but they pose an entirely different energy-security problem because of safety concerns that have persisted since the Chernobyl disaster (and that have been intensified by perceived vulnerabilities to terrorist attacks) and because of the difficulty of safely burying radioactive waste that remains lethal for thousands of years.

At a U.S. Steel plant in Gary, Indiana, and in steel plants all over the world, a byproduct of the iron-smelting process is "blast-furnace gas," which consists mostly of carbon monoxide and nitrogen, with some hydrogen and carbon dioxide. The monoxide and hydrogen make it flammable (and toxic), so it must be flared if a beneficial use cannot be found. But in this plant, the blast-furnace gas is captured to produce steam, which drives a steam turbine powering a generator with an annual output of 100MW—even more than at the Mittal coking plant a few miles to the west.

A fourth kind of waste-energy stream is produced by decompression. About 8 percent of the natural gas shipped by pipeline is used for compression of the gas itself, to drive it through the pipelines. At the delivery point, this compression energy is lost. Yet a simple back-pressure turbine, costing a few hundred dollars per kilowatt, can convert that pressure to useful electricity. This process alone could add another 6,500MW of carbon-free electricity in the United States, saving roughly 1 percent of U.S. fossil-fuel consumption and the greenhouse gas emissions.

The Biggest Energy Drain

A fifth, and very different, waste-energy stream is *low-temperature* heat, which is dumped into the air or water in enormous quantities by—of all people—the big centralized electric utilities. You might wonder, why would a company that's in the business of selling energy *dump* energy? The reason is that, unlike high-temperature waste heat, low-temperature heat can't be used to make electricity, so the central power plants just blow it into the sky or into a nearby river or pond.

However, that doesn't mean that low-temperature heat can't be used. It's just that it can't be used in the places where most central power plants are located—far from the cities or towns they serve. Although electricity can be transmitted many miles over wires, hot air or hot water can't be sent any distance without cooling. But if the heat can be used just a short distance from the power plant, it has an immediate energy-saving and carbon dioxide–reducing benefit.

The main use of low-temperature heat is for warming homes or buildings. In most U.S. communities, space heating is provided by burning oil, natural gas, or propane, or by purchasing electricity from a utility that burns coal or natural gas. In other words, nearly all U.S. homes and buildings (with a few solar or wood-burning exceptions) are heated by fossil-fuel combustion, directly or indirectly. If they could be heated by the low-temperature waste heat from electric-power production, the fossil-fuel combustion now used to produce that heat could be completely eliminated.

The potential for that saving is enormous. With the conventional U.S. electric-power system operating at an average efficiency of just 33 percent (including transmission losses), only one of every three units of energy going into those plants ends up being delivered to consumers in the form of electricity. The other two units are discarded as waste heat. The obvious question is, how can we get the heat to where it can be used?

One answer is a strategy called CHP. Among energy experts, CHP refers not to the California Highway Patrol, but to something that could make an arrest with more stopping power than most policemen ever get to use: It could arrest what is perhaps the single largest of the many leaks that are draining America's energy supply. CHP is the strategy of combined heat and power—producing both heat and power in the same plant as saleable products. Because a conventional electric-power plant produces only electric power in a facility that is typically located in the boondocks, the power must be transmitted over costly (and ugly) power lines to cities. But suppose the power is generated right in the basements (or rooftops) of apartment houses, shopping districts, university campuses, or industrial parks where it is needed, and where the buildings can then use the waste heat for heating and hot water. That kind of system, called *decentralized* CHP, or DCHP, eliminates not just the financial and environmental costs of buying power from so-called central plants, but also the substantial costs of transmitting power at high voltages over long distances.

The Mittal and Kodak cases previously described are limited forms of CHP because both heat and power are produced and used. Unfortunately, it's not as easy to cite current examples of DCHP used for a shopping mall or office park in the United States because DCHP is, for most purposes, illegal in every one of the 50 states. You can generate electricity for your own use or sell it back to your utility monopoly (at a price it decides), but you can't sell it to your neighbors. It is actually illegal in every state to send electricity through a private wire across a public street. That's why Mittal Steel can't sell clean, cheap electricity to its neighbors in East Chicago, for example. We return to this topic in Chapter 5, but for now let's just say that the laws blocking DCHP need to be modified. If politicians and policymakers are serious about energy independence, the laws that created those utility monopolies in the 1920s, under very different circumstances, will have to be changed.

DCHP is now used routinely in other countries. In much of Europe, a form of CHP called "district heating" has been implemented for decades. Otherwise-waste heat from local power generation is distributed short distances through pipes to nearby users (typically, apartment buildings). It saves fossil fuel that would otherwise be burned just to produce heat, and it replaces the highly inefficient system of conventional space heating by means of a furnace in the basement. But it's practical only in very densely developed areas with power plants nearby. District heating isn't of much use in the United States, with our sprawling cities, suburbs, exurbs, and scattered small towns.

More advanced systems of DCHP, in which gas turbines or diesel engines (or eventually high-temperature fuel cells) generate both heat and power in the same building, are already up to speed in some of the more technologically advanced countries. CHP accounts for more than 50 percent of the electricity produced in Denmark, 39 percent in the Netherlands, 37 percent in Finland, and 18 percent in China. Governments have achieved these results mainly by requiring the utilities to reduce carbon emissions and, consequently, find markets for their heat, which has meant locating new electricity generation right in the places where heating is needed.

Not incidentally, those countries (except China) rank among the top countries in living standards—places where in-house power generation would be unacceptable if it weren't unobtrusive, quiet, and clean. If power and heat could be cogenerated in individual buildings in the United States *while retaining connections to the grid*, virtually all new additional capacity could be decentralized. That possibility is not a sci-fi dream; it's a right-now reality. Only the laws protecting utility monopolies need to be changed. One of the rules of this book is that everything we propose for the next decade (and most of what we forecast for beyond) can be achieved with existing technology—already in use somewhere in the world—using domestic energy resources. In a 2008 report, the International Energy Agency (IEA) projected that if future demands for new capacity were to be met by adopting CHP, but without significant changes in the laws, global savings in capital costs would be $795 billion. Given the current U.S. share of global energy consumption, the U.S. share of that saving could be between $100 billion and $200 billion. We think the actual potential is higher.

The initial reaction of many legislators, bureaucrats, economic advisors, utility commissioners, and city planners to this idea of decentralized heat and power might be quick dismissal, due to that pervasive belief that central power plants are optimally efficient and that small-scale production could never compete. In the early twentieth century, that was true—and that's how utilities got the legally protected monopolies they now control. But while central power plants have not significantly improved the efficiency with which they generate and deliver power in 40 years, small systems have improved dramatically. Small gas turbines and diesel engines today are almost as efficient for electricity generation as the large steam-generating systems in the central power plants, especially when transmission and distribution losses are taken into account. And when the potential for local use of waste heat is added in, they are far *more* efficient because they eliminate much of the need for fuel currently burned for space heating and water heating.

End-Use Inefficiency Shock

To put the potential of combined heat and power in perspective, it's important to remember that when we say the efficiency of the present electric utility system is 33 percent, that's just the efficiency with which it generates and delivers power to the consumers. Only one-third of the energy contained in a barrel of petroleum or oil-equivalent ends up as electric energy arriving at the meter. To calculate the overall efficiency of the actual energy *service* (lighting, heating, and so on), you need to multiply that 33 percent efficiency by the efficiency with which the consumer *uses* that delivered power, whether it's to run a motor or to turn on a light.

Everyone knows now that an incandescent light bulb (that familiar "bright idea" symbol of the past century) has very poor end-use efficiency in terms of lumens per watt, and that compact fluorescents are far better. But although fluorescent lighting gets about three times the efficiency of incandescent (about 15 percent versus 5 percent), when multiplied by the 33 percent of the power delivered to it (.33 × .15), the total efficiency of the compact fluorescent light is still just 5 percent.

Similarly, we might be encouraged by the advent of plug-in electric cars, but although the average mechanical efficiency of an electric motor is between 60 and 95 percent (depending on size, speed, and so on), the charge–discharge cycle for the battery itself loses about 20 percent each way (in and out). A car using plug-in electricity from a 33 percent–efficiency central power plant might have an overall efficiency of power to the wheel of 16–18 percent. That's more efficient than a conventional gasoline-powered vehicle operating in city traffic, but it still wastes the energy embodied in more than five of every six barrels of oil-equivalent.

Then consider the *payload* efficiency you get when you drive a car. Set aside the question of whether it makes sense, in a country where energy is no longer cheap, to move more than a ton of steel, glass, and rubber (plus fuel in the tank) to transport your 200 pounds, or whatever you and your briefcase or shopping bags weigh. If the efficiency of moving the car itself is 10 percent—typical in the United States—the payload efficiency of what's being transported (assuming it's one-tenth the weight of the car) is a tenth of that, or about 1 percent. If you carry a second person, or have a lot of luggage, the payload efficiency might be 2 or 3 percent. If the car is hybrid or electric, you might get up to 4 percent. For stationary uses, a comparable inefficiency prevails. Someday historians will shake their heads in wonder.

If you add up all the different kinds of energy use in the United States, the overall efficiency just for producing useful work is currently around 13 percent (and that's before taking payload inefficiency into account). It's as if a father goes out to buy seven ice cream cones for his kid's birthday party, and six of them fall on the ground as he's walking out of the store. The bad news is that a lot of ice cream is lost. The good news is that the dad's dexterity has lots of room for improvement.

When President George W. Bush and his would-be successor John McCain urged America to address its energy independence problem by drilling more holes in the ocean floor, they might not have been aware that they were recommending a course of action that would do nothing to improve the country's truly crippling energy inefficiency— nothing to relieve either near-term dependence on Middle Eastern oil or the longer-term problem of global warming. If the country were to adopt then–Vice President Dick Cheney's nightmarish scheme to build 1,300 new coal-fired central power plants, the effects would be even more devastating: The energy efficiency of the country would be barely on life support, and carbon dioxide emissions would rise to even more dangerous levels. Or if we followed the "clean-coal" route being promoted by the coal lobbyists and utilities, the carbon dioxide would continue to climb at approximately the same rate, but the cost of power would rise sharply—and the economy would be further crippled. ("Clean coal" might sound reasonable to people who don't get a kick out of oxymorons, but the process of converting the coal to gas— which gets rid of the fly ash and sulfur—makes the coal energy about twice as costly to deliver, and the process of capturing and storing the carbon dioxide produced by combustion doubles the cost again.)

Suppose that, instead of following the reflexive impulses of politicians pandering to an electorate fearing for its energy security, the Obama administration were to systematically put together a strategy combining just the two major opportunities outlined in this chapter: (1) Recycle high-temperature waste heat, steam, or flare gas in industrial plants, and (2) encourage the shift of mainstream electric-power production from centralized to decentralized heat-and-power production. How much would the country's need for fossil fuels be reduced, and how far would that take us toward full energy independence?

First look at recycling waste energy. We noted that the U.S. Steel plant in Gary, Indiana, produced about 100MW in 2004, and Mittal's Cokenergy plant produced 90MW. Approximately 1,000 other U.S. plants are already doing waste-energy recycling. Most of them are smaller than the Indiana giants, but together they were contributing 10,000MW of electric power per year to the national total, according to the latest available data. Yet according to a recent study for the U.S.

Environmental Protection Agency, 19 different U.S. industries could have profitably generated more than 10 times that amount by recycling wasted heat. Even accepting a more conservative estimate by the Department of Energy, the profitable potential for energy recycling is six or seven times greater than the current level of recycling. Most of it would be clean electricity replacing power currently purchased from coal- or natural gas–burning utilities.

The approximate capacity of conventional (fossil fuel–burning) power plants in the United States in 2007 was 900,000MW, or 900 gigawatts (GW). The installed capacity of waste-energy stream recycling was 10GW. And the solar-photovoltaic (PV) capacity was 0.1GW. By 2009, PV had grown to nearly 0.2GW, and President Obama projected that the solar energy industry would double again in the following three years. As an industry grows larger, it's unrealistic to expect it to continue expanding at the same rate, but suppose the solar-PV industry continued doubling every three years. That would bring it to roughly 1GW by 2015—still just a fraction of 1 percent of U.S. electricity production. But in the meantime, if energy waste-stream recycling doubled at the same rate, it would reach 40GW—with more room to grow. If the *full* potential of energy recycling is exploited, we can generate up to 10 percent of U.S. electricity without generating carbon emissions or burning any additional fossil fuel. Granted that solar-PV is the golden future and cleaning up dirty fossil fuel is the prosaic present, a hard reality in the present business climate is investment cost. And the reality is that the waste-energy recycling option is much cheaper.

For wind power, the near-term prospects are stronger, but not yet strong enough. U.S. wind capacity reached 0.8GW in 2006, and wind is economically competitive on a per-kilowatt basis in some locations. But the actual output of wind facilities is intermittent, so the real output is less than that of a plant that is operating continuously. Even assuming a very optimistic growth trajectory for wind power, the recycling of waste streams from aging fossil coal– or natural gas–burning facilities will have greater potential for affordable carbon-free power, at least until 2013. Beyond that, the capacity for solar and wind power to continue growing geometrically becomes unrealistic.[4] But even at the most rapid continued growth conceivable, it would be many years before solar-PV and wind power could replace more than half of the nation's fossil-fuel power. To keep the economy adequately functioning for that time and beyond while continuing to reduce carbon dioxide emissions, large investments in renewables must be joined by equally large (and initially more productive) investments in energy recycling.

Then consider the central power plants and the potential for ramping up U.S. power production by phasing out of "centralized" into decentralized CHP. Approximately 3,855 utility-owned or municipal electricity–only power plants currently exist in the United States. Studies by energy engineers show that the 33 percent efficiency of those plants plus their massive transmission and distribution infrastructure could be increased to around 60 percent efficiency if all new and replacement capacity were decentralized. That shift could take many years, but if the laws that prevent it were changed quickly, a substantial bump in electric power production—while achieving a net *reduction* in fossil-fuel use—could be achieved within a few years. If no new central plants are built and half of the old ones are phased out and replaced by CHP, half of the industry's 900GW capacity could shift from 33 to 60 percent efficiency—increasing total U.S. electric power by roughly one-third, while cutting emissions by one-third and using no additional fossil fuel.

That two-part strategy—recycling industrial waste energy and beginning to decentralize electric power generation—would constitute a huge stride toward energy independence *and* toward the parallel goal of sharply cutting carbon emissions. But it's not the whole story by a long shot; it's just the first chapter after the wake-up call.

In comparing the strategy we've just outlined with the option of drilling for more oil off the American coasts, we like to use an analogy. Suppose you have a farm in upstate New York where you keep seven wild mustangs in a corral. One day you discover that six of them have

---

[4] Suppose you decide to save a penny today and double the savings every day. Keep doubling every day by working hard, and you'll have $20 million in a month, but you probably can't keep it up for more than the first ten or twelve days.

escaped. Do you immediately plan a costly new expedition to find replacements in wild-horse country 2,000 miles away, or do you try to retrieve the ones that escaped into neighboring fields and can't have gone far? Thinking of horses as units of potential work (horsepower-hours), and keeping in mind that six of every seven units of U.S. energy extracted from coal mines or oil wells escape before they can produce useful work or heat, doesn't the same question about retrieval versus replacement apply? It will be vastly cheaper to retrieve a barrel's worth of energy from a waste-energy stream that already exists in Allentown, Pennsylvania, and use it for electricity that's needed right there, than to retrieve that barrel's worth from a hole a mile deep under the Pacific Ocean off the coast of Santa Barbara and then refine it and ship it 3,000 miles.

To take this analogy one step further, we like to recall that before European explorers arrived in North America, no horses lived there. Horses originated in Central Asia and the Middle East. Later, those formerly Arabian imports became an indispensable part of the pioneer American culture and economy. Now as energy pioneers of the twenty-first century, we have an opportunity to do the same with horse*power.* Our dependence on Saudi Arabia for the energy needed to run a modern economy can come to an end. *We already have the horsepower in our own country.* Like the farmers, mail carriers, and cowboys of an earlier time, we just need to learn how to harness it.

3

Engineering an Economic Bridge

For those who had enough vision to foresee the economic tsunami of climate change we now face, it was only natural to also foresee that alternative energy technologies could eventually provide the economic boost needed to replace fossil fuels. But the key word here is *eventually*. Many visionaries didn't realize that, in the fossil fuel–dominated economy we're stuck with for at least the next quarter-century, it isn't the burgeoning of new industry that drives growth, so much as how the *existing* technology is used. In political speech, the image of an old coal-burning steam boiler can't compete with a shining new solar farm for inspirational effect. But in the real world of industry, if the right modifications are made, the old coal-burner might provide access to more new carbon-free energy per dollar of investment than the solar farm can in the next decade.

The post–fossil fuel economy that has energized progressive thinkers and organizations such as Al Gore's Wecansolveit.org is an economy we need to be building now with all possible speed. But even under the best of circumstances, it will require time-consuming research and development—and massive infusions of capital—to complete. Building the energy bridge we propose won't require as much new capital. But to get to the post–fossil fuel economy without sinking into Third World decay in the meantime, we need to use old technology in new ways. Investors and policymakers need to give major attention to getting more energy service per unit of coal, oil, and natural gas *as well as* bringing the new energy industries up to speed as fast as possible. (Fossil-fuel interests might find it irresistible to quote the first half of that last sentence out of context. But let us emphasize that the only reason for that "major attention" to using fossil fuels efficiently is to safely get *free* of them.)

Our reason for describing the energy policy of the next decade as a "bridge" is the problem of cost. When the second Bush administration decided to give the economy a prod by mailing free money to everyone in early 2008, it was presumably to provide a bridge to the day when a revived consumer-spending economy could begin paying back the cost of the stimulus.[1] The economy was not visibly stimulated. A few months later, when a capital transfusion was needed for the whole financial system, the same assurance was implicit: The money would be paid back when the economy resumed its robust growth, not *if* it resumed. As Bush left office and the crisis continued to worsen, the new Obama administration upped the ante with its nearly trillion-dollar American Recovery and Investment Act of 2009. By then, the country was in a recession so deep that it was unseemly to question whether this new expansion of the national debt would produce the anticipated rebound. The situation was so dire that officials discreetly refrained from publicly discussing how much *additional* cost would be exacted in the coming years by (1) the soon-to-be-spiking price of oil when the global supply begins to decline even as demand from China's gargantuan, car-hungry population continues to rise, and (2) the costs of escalating climate-change damage.

Alarmingly, economists who have studied the likely impacts of climate change have nearly all agreed that those costs will be substantial, cutting GDP growth by 0.1–1.5 percent, depending on model assumptions. Traditional economists and physical scientists sharply disagree about how those costs will—or should—be incurred. Conservative economists who believe that the economy will soon resume vigorous growth—and make our grandchildren richer than we are—have argued that the present value of climate action should be sharply discounted because the money is worth more to us now than it will be to those rich kids later. They say it's better to have the money available to spend or invest than to have it tied up in programs that might not yield benefits for decades. In keeping with the still-prevailing

---

[1] A headline on Forbes.com in February 2008: "White House Report Says Economy Will Rebound in 2008." A headline from Reuters news service in July 2008, quoting Treasury Secretary Henry Paulson: "Economy Needs Months to Recover: Paulson."

economic ideology, they also argue that actions taken now by government-mandate s will hamper economic growth, and that the resulting loss of GDP might be a greater burden to us in the long term than the cost of climate damage.

Climate scientists and environmentalists argue just the opposite—that the future cost of climate damage will be enormous, and that immediate action is needed to offset it. But both camps agree that society will bear a net cost. Experts disagree about how much cost we should incur now versus later, but few people entertain the hope that we can get through the next decade or two unscathed.

The authors of this book have a different view. We have no reason to question the Intergovernmental Panel on Climate Change (IPCC) projections of staggering physical damages; they represent the consensus of the leading climate scientists from every industrialized country on Earth. Coastal cities from Shanghai, Canton, and Dhaka, to New York, Houston, and Miami—cities with populations up to 20 times greater than that of New Orleans—are literally in danger of annihilation, if not next year, sometime within this century. Low-lying island groups such as the Maldives, the Andamans, and much of Micronesia are facing almost certain extinction. Food and fresh water supplies for much of the global population will be at risk. The one–two punches of destruction and disease could drive unprecedented refugee flows, resulting in widespread disorder and conflict. However, the worst of those damages are likely to come after the transition period (the bridge over the economic chasm). The real question now is how much it will cost *in the next few years*—as we begin the transition—to take mitigating action.

We disagree with the neoclassical economic modelers (such as William Nordhaus of Yale) who say that any action such as government-mandated carbon emissions limits will cause a "deadweight loss" that reduces GDP by constraining the "optimal-growth" path. We disagree with those economists (such as Nicholas Stern of the London School of Economics) who seem resigned to the idea that climate action might require a net loss of perhaps 1 percent or more of GDP per year. As we briefly discussed in the Introduction to this book, we think that the mathematically convenient assumption of the economy being on an optimal path is inconsistent with many facts.

We think smart government interventions at this point might eliminate some of the barriers—such as monopolies and "lock-ins"—and
bring us closer to optimality, not farther away. Our view is that
although we could incur significant administrative costs in such things
as running a redesigned carbon-trading system, and significant capital
costs in shifting auto production to higher fuel economy, the strategies that economists have *not discussed* require very little up-front
costs and can achieve savings that will more than offset those costs. A
little later, we explain why. First, we want to look at what the reengi-
neering strategies must be, if the United States is to cross the transitional bridge safely.

Double Dividends

The "me-oriented" view of financial well-being and security
embraced by many Americans (not just celebrities and hedge fund
managers) during much of the past 30 years defines economic benefits mainly in terms of individual income and wealth. In the corporate
world, that has meant gains mainly for top executives. In the general
population, it has meant the promises of the government taking less
money from your pocket.[2] Looming problems arise from this focus on
personal reward, about which commentators have written compelling
critiques. But with a few notable exceptions, such as President
Obama's scolding of Wall Street executives for taking huge bonuses—
and effectively rewarding failure—at a time when the national economy was on its knees, some of the most important problems with (and
critiques of) this me-oriented view have been given only superficial
attention by mainstream media and politicians.

For example, there is the problem of what economists call
"externalities"—costs that are paid by neither the buyer nor the
seller in a transaction, but that have to be taken from *someone*,
whether it be a third party who has no interest in the transaction or

---

[2] In 2006, the last year before the housing crash, average pre-tax income for the
top 1 percent of Americans grew by an average of about $60,000 per household;
for the bottom 90 percent of households, it grew by just $430. As the economy
began its protracted plunge in 2007, the gains of the top few rose even more, but
the incomes of the majority fell.

a person who has not yet been born and can't speak up. For example, the cost of cancer treatment for a person who has breathed diesel truck exhaust containing the chemical 3-nitrobenzathrone is an externality that isn't paid for by either the company making the diesel fuel or the trucking company buying it at the pump. Yet that medical cost is counted as a part of the gross domestic product (GDP), so it is treated (in the national accounting) as a good thing. Similarly, the cost of cleaning up an oil spill, or the price a criminal pays for a gun, adds to the GDP. The problem is that although a few individuals gain from the transaction, many others are worse off in the end. Higher GDP always counts as a gain to the national economy. We, along with quite a few economists, believe it's time to rethink the use of that traditional measure of wellbeing.

Then we have the problem of discounting the value of future benefits in favor of immediate gratifications. Many studies have shown that, for most people, immediate gratifications trump benefits that are more distant in space or time, or more hidden from view. The home-theater TV you can buy on sale today, or that weekend in Vegas, might take preference over the month of your child's college education that money would buy ten years from now. The office seeker who promises to cut your taxes will all too likely get your vote over the one who plans to use those taxes to repair deteriorating water mains or clean up a contaminated aquifer that you can't see.

The "me-first" view tends to discount societal benefits in favor of individual ones. A pop-psychology book that encourages you to "be good to yourself," "believe in yourself," or learn that you must "love yourself before you can love others" will probably sell a lot more copies than a book that suggests making sacrifices for the benefits of people not yet born. A woman who is outraged by a spike in the price of gasoline might have little interest in the rate of carbon dioxide emissions from the nation's cars, even though that might ultimately have far more impact on her health and wealth than the price of gas ever will. A man who is preoccupied with protecting his personal wealth might have little interest in the national debt, even though it might profoundly threaten his long-term security.

If we take a somewhat broader view of economic well-being, it needs to include social benefits as well as individual ones, benefits of cooperation as well as competition, and national or community security as well as personal security. That broader view is essential if we hope to build real protection from climate change, because such protection requires wide cooperation on an unprecedented scale. Forests burning in the Amazon jeopardize the future of American cities as much as gangs or terrorists do, if not more so. A politician who pontificates that money spent on critical public projects is "the government taking money from your pocket" is not just a demagogue—he's a con man. He might just as well tell you that money spent on the police department payroll is money taken from your pocket—an assertion he might have second thoughts about when a mugger takes the wallet from *his* pocket and no policeman is around.

Conservatives have often fended off expenditures on societal benefits by suggesting that individual and social benefits are mutually exclusive—that adding to one subtracts from the other. For example, extractive industries (mining, drilling, and logging) widely promulgated this "zero sum" view during the early years of the environmental movement by claiming that expenditures on environmental protection were bad for the economy. In the Pacific Northwest, where the spotted owl became a symbol of efforts to save forests from decimation by loggers, bumper stickers appeared with the words "Kill an Owl, Save a Job."

Abundant analysis since then has confirmed a very different proposition: Protecting the environment (including the climate) not only doesn't harm the economy; it is essential to the economy's capacity to function *at all*. For any individual American or business, the prospect of surviving the next decade without bankruptcy or impoverishment requires preserving the profit motive but significantly strengthening government oversight. As individuals, we naturally covet our opportunities for personal achievement, security, self-esteem, and pursuit of happiness. But for that individual pursuit of liberty and happiness to be even possible, we need communal cooperation—protection from robbers, toxic pollution, environmental ruin, predatory lending, or climatic catastrophe.

It's in this context that we have identified a range of opportunities for actions that produce what some analysts call "free lunches," although we also use the term "double dividends"—actions that provide the societal benefit of climate mitigation *and* the private financial benefits of lower costs (or increased business income and profit). In economic terms, some of these actions will actually incur "negative costs"—they will reduce carbon emissions in a way that can actually increase a company's profits and the nation's GDP. They will boost the nation's economy in a far more targeted way than a tax cut or a something-for-everyone "stimulus" package possibly can.

Not all the climate and energy actions aimed at reducing global warming emissions and increasing U.S. energy independence will produce double dividends, but a surprising share of them will. And when all these actions are taken together, we believe that the overall economic cost will be far less than the major studies to date have projected. Perhaps the net costs will even be less than zero when all the benefits are taken into account, although much depends on the details. (For example, the administrative cost of a national or global carbon-trading system could be either enormous or very modest, as we discuss in Chapter 10, "Policy Priorities.") We think that large parts of the proposed strategy, starting with the energy-recycling opportunities mentioned in Chapter 2, "Recapturing Lost Energy," can be implemented with double dividends within the very short payback time needed to be of value in building a transitional bridge.

The need for short-term payoffs is a critical consideration, particularly in the wake of the very vague and often misleading claims made during the 2008 presidential campaigns. When John McCain advocated an intensified program of drilling for oil and the construction of scores of new nuclear and "clean-coal" plants, he was invoking actions that would require 10–20 years to produce any additional energy. And the proposed program would only prolong (and, thereby, worsen) the nation's dependence on fossil fuels by massively investing in new fossil-fuel infrastructure that would take another generation to depreciate and replace. Barack Obama's stated plans seemed more informed, but still reliant on strategies that would be too slow, too costly, and ultimately too ineffective to get us across the chasm. The actions we list in this next section—and describe in detail in the coming chapters—don't take as long to implement, and begin to phase down some of the fossil-fuel infrastructure instead of bulking it up.

Main Girders of the Energy-Transition Bridge

We envision a national energy strategy that entails eight main components. Of these, *one* (increased energy efficiency of consumer products, including automobiles) has received serious attention in the media. Two others (increased energy efficiency in buildings and industrial plants, and decentralized electric power) have received moderate attention in the industrial world and technical or academic literature, but very little public discussion. Very few people know the other five. Yet all eight of these components have proved productive in real-world applications. None are the kinds of speculative ideas (such as lunar energy production or cold nuclear fusion) that may or may not pay off at some future time. Only two of the eight require significant research and development (R&D), and even they are ready for immediate use in the meantime. All but one can begin paying off in the next few years. The eight components, with cross-references to our more detailed discussions of them in this book, are the following:

1. **Recycling waste-energy streams**—In Chapter 2, we mentioned the Cokenergy plant in East Chicago, Indiana, which annually produces 90 megawatts (MW) of bonus emissions-free electricity from its waste heat. About a thousand U.S. plants generate power this way, producing about 10,000MW. But as we noted, there is untapped potential for *ten times* the amount currently recovered that could be harnessed within the next few years, with no added fossil fuel use and, in most cases, no carbon emissions. This well-proven approach can produce up to 10 percent of U.S. electric power generation without fossil fuel combustion.

2. **Utilizing combined heat and power (CHP)**—We also introduced this girder in Chapter 2 and noted that most of the potential is still untapped. In the present electric-power system, two-thirds of the energy going into the system is wasted as low-grade heat. If we produce power in ways that capture and use that heat, we can eliminate much of the fossil fuel currently burned to heat houses and buildings. In contrast to other girders, such as end-use efficiency, this is an option virtually never

discussed by elected officials. It's one of the most politically daunting actions we propose, but the double-dividend benefits (for nearly all users of electricity and all regular breathers of air) would be so great that political leaders who view the importance of the transitional bridge with clear eyes will need to bring this taboo subject into imminent consideration. We discuss this further in Chapter 5, "The Future of Electric Power."

3. **Increasing energy efficiency in industrial processes and buildings**—Hundreds of U.S. companies have reduced energy use (and carbon emissions) by making engineering changes in their processes or equipment. Some of the improvements produce recoverable forms of energy, as previously described, but many others cut primary energy or, most often, electricity consumption. Yet these opportunities have been widely ignored because of two mistaken but pervasive beliefs in business economics: (1) the easy savings have already been made, and (2) further conservation efforts aren't worth it. In Chapter 4, "The Invisible-Energy Revolution," we show that such investment is unquestionably worth it. In many cases, returns on investment have been quick and lucrative. Enough double-dividend opportunities exist to help span the transitional bridge for the next 10–20 years.

4. **Increasing energy efficiency in consumer end uses**—This is the one girder of the bridge strategy that has been well publicized and has already taken us an encouraging step in the direction we need to go. Compact fluorescent lights, hybrid cars, energy-efficient appliances, multipane windows, and so on have gone mainstream. They haven't yet noticeably slowed the global output of carbon emissions, although that observation doesn't take into account what the emissions would be by now if those improvements hadn't been made. But a large share of the national potential still remains untapped. Because consumer efforts have been widely publicized in other media, we touch on most of them fairly briefly—mainly to confirm their importance in raising public awareness that other girders are also needed for the bridge to work. It's important to emphasize that changing consumer behavior in this domain is important (see Chapters 6–8), but it's still just a beginning.

5. **Kick-starting the micropower revolution, or "rooftop" revolution**—Breaking the utility monopolies on the distribution of electric power to further the development of combined heat and power will be a key underpinning of this girder. But beyond that will come a more fundamental separation of powers between the big public functions of utilities and the small local ones of private homes and small businesses. Small-scale production of power in individual homes, offices, and even vehicles or boats might seem farfetched today, but recall that the idea of a computer in every home or office seemed equally remote a few decades ago. Today's great central power plants, with their energy-leaking power lines marching over our national forests and farm fields, are similar to the giant mainframe computers of the 1960s. As we discuss in Chapter 5, electric power generation is on the verge of a revolution.

6. **Substituting energy services for products**—In the real energy economy, it isn't coal, oil, or gas that Americans, Europeans, or Asians have an appetite for; it's light, heat, mobility, communication, and entertainment. It's not the energy itself that we need, but the services that energy provides. For example, few people need gasoline, but nearly everyone needs mobility. In some cases, the services we demand can be provided in ways that make little or no direct use of fossil fuels. We discuss this in Chapters 8 and 10.

7. **Redesigning buildings and cities for climate change**—This girder will take a long time to build, but getting started soon is vital. It's already possible to build houses that require only a small fraction of the energy that the average new home consumes. Already, thousands do so in Germany and other industrialized countries. In time, urban dwellers will also be able to sharply reduce the amount of energy they expend on transportation, while actually improving mobility. (See also girder 4.) We also need to prepare for the hard reality that rising sea levels and growing storm-surge risks will soon force hundreds of the world's coastal and river delta or riverside cities to begin relocating their low-lying districts to higher ground. Massive redevelopment during the next half century will provide obvious opportunities for more compact urban

design, thereby improving both energy efficiency and the quality of life (see Chapter 8).

8. **Reforming fresh-water management strategies**—Current water-delivery projects use prodigious amounts of energy for pumping water long distances, up hills, or out of deep wells. Water management has become a major user of energy, and alternative strategies for water supply, which weren't considered when the infrastructure was built at a time of low energy costs, can now provide opportunities for dramatic reductions in fuel use (see Chapter 9).

Why the Taboos?

This combined lineup (or something close to it) is not only capable of providing the needed transitional bridge, but is probably the only safe strategy for doing so. Yet as we noted, most of the essential girders rarely become topics of public discourse. During the 2008 presidential campaigns, Senator McCain didn't mention any of them in his stump speeches or debates; all his remarks about energy concerned new sources of supply. Senator Obama mentioned only a few of them, notably end-use efficiency in cars. To their credit, both campaigns signaled a break from the nightmarish energy policy of Vice President Dick Cheney, who scorned efficiency and shrugged off climate change. But both campaigns reflected the general media emphasis on new supply, updated only by an inspirational call for wind and solar power to provide much of that new supply. Neither man articulated the reality that the nation needs to bridge a deep supply chasm. And after Obama took office, we saw few signs—other than the allocation of some stimulus money for energy efficiency—that the barriers and blockages in the existing system were to be reconsidered.

The mainstream public discussion of the coming climate and energy crisis incorporated several large blind spots, or even taboos:

- No one acknowledged that *the amount of primary supply is not the same as the amount of energy service.* The amount of useful work produced by a barrel of oil can be very little or a lot.

- No one pointed out that *an archaic set of laws perpetuates an outmoded electric power system,* which, if reformed, could greatly increase its output while reducing its fuel use and emissions.

- No one called attention to the impending challenge of "peak oil." Although several popular books (and many academic ones) have focused on this topic, and energy experts and blogs discuss it extensively, the media during the early stages of the economic crisis did not—even though peak oil will likely begin the onset of painful shortages, price gouging, and widespread economic and social disruptions as soon as economic growth begins again.

- Finally, no one suggested that the economic downturn which began in 2008 was fundamentally different than previous ones—that we were indeed at the brink of a chasm, and that *we must build a bridge from something other than new supplies of oil or gas.*

Why the taboos? It's not a mystery; a combination of unspoken political expediency on the part of the media (most of which are heavily dependent on the advertising of motor vehicles, real estate, and other fuel-dependent industries) and technical complexities of our economic system has squelched the discussion of these uncomfortable facts. Any politician who is willing to buck the interests of the oil, gas, coal, petrochemical, and electric utility industries—and fossil fuel–dependent businesses such as motor vehicles, airlines, and agribusiness—risks cutting off the financial lifelines on which elected officials depend. By 2008, it was relatively safe to call for increased investment in wind and solar energy, but not safe to attack the absurdities of car-dominated transportation or electric utility monopolies.

As for the nondiscussion of such huge opportunities as increased energy efficiency in industrial processes, the issues are simply too complex to be reduced to sound bites. Materials processing and manufacturing industries utilize millions of different energy-consuming operations, most of them invisible to the public. In the 2008 presidential campaign, John McCain tried to tap the working-class vote by invoking the job of "Joe the Plumber." Everyone knows what a plumber does. We might not be so familiar with the jobs of the millions of men and women who work with the tens of thousands of industrial processes that consume fossil-fuel energy and emit global-warming gases behind windowless walls. Most of these operations were designed and set up at a time when energy was cheap and not a primary factor in business management. We can reconfigure or reengineer the vast majority of these operations to cut energy use and emissions, but managers have only belatedly begun to see what enormous potential is there.

4

The Invisible-Energy Revolution

In June 2008, a few months before the financial-sector train wreck, the U.S. media were roiling with news about skyrocketing energy prices and the widening global gap between supply and demand. Gasoline prices were at the top of the TV news every night, sporadically interrupted by video of the latest weather catastrophes in the Midwest. Bush administration officials spoke with undisguised anger about OPEC's lack of attention to U.S. needs.[1] In a TV interview, conservative news host Sean Hannity told liberal Sen. Barbara Boxer that high gas prices were causing Americans terrible hardships and that caregivers were unable to get to the people they cared for. Boxer responded that the Bush administration had been driving the economy into ruin for seven and a half years and that this was one of the results. Hannity countered, "But what would *you* do?" She didn't quite answer the question. The Republican administration's energy strategy was clearly a disaster, but it wasn't clear that the Democrats were offering much of an alternative.

In this inane exchange, neither the politician nor the pundit recognized the possibility that if the country could find some way to get more energy service per barrel of the oil it was already consuming, it might not *need* more supply. If we could find a way to double the

[1] President Bush, who had scornfully criticized Barack Obama for saying that, if elected, he would be willing to hold talks with foreign leaders who are "tyrants," flew to Saudi Arabia that month to talk with the Saudi King Abdullah—a well-known autocrat, if not a tyrant—about pumping more oil for the United States. A news video showed a smirking, red-faced Bush swaying arm-in-arm with the king in a ceremonial dance, looking as embarrassed as a 12-year-old boy who'd been asked to appear on stage in a dress. His apparent message: For more oil, we do what we must.

energy efficiency of our economy, the effect on economic output would be equal to doubling the nation's energy supply. Efficiency doesn't get headlines. Yet a month before the Hannity–Boxer interview, on June 15, the American Council for an Energy-Efficient Economy (ACEEE) issued an analysis showing that, although U.S. energy-service consumption had increased greatly during the previous 38 years, *energy-efficiency improvements provided three-fourths of the increase in consumption and new supply provided only one-fourth.* "It's the U.S. energy boom that no one knows about," said the ACEEE's director of economic analysis, John Laitner. In their exasperated debate, Hannity or Boxer could have thought to ask whether consumption could actually be *cut* by increasing energy efficiency. That's a critical question, but they didn't ask it.

Perhaps the media didn't respond to the ACEEE report because they were confused. For decades, an established trend of the American economy has been a long-term decrease in energy *intensity*—the amount of energy needed to produce a dollar of output keeps declining. In its press release, the ACEEE noted that during the previous 38 years, energy consumption per dollar of GDP had fallen dramatically, from 18,000 British thermal units (Btus) per dollar in 1970 to just 8,900Btus in 2008. Greater energy efficiency is one of the causes of decreasing intensity (less energy input per unit of work helps to produce more economic output per unit of energy). But other activities, including exporting energy-intensive industries such as aluminum smelting, also cause declining intensity. Energy intensity can fall even as the total amount of energy used grows, if GDP grows even faster. Reporters might have seen this as a nonstory.

What the media missed was the sheer *size* of the efficiency component—the three-quarters share. By 2008, Bush–Cheney officials had so repeatedly dismissed energy efficiency as having only marginal value, compared with new supply, that the real news became what journalists call a "buried lead"—a story whose most important information lies somewhere in the seventeenth paragraph, which most readers overlook. In past studies of U.S. energy trends, the role of efficiency had never been so clearly separated from intensity. Now the ACEEE had provided what it described as "the first attempt to quantify the overall impact of the hidden U.S. energy-efficiency boom." The report, titled *The Size of the U.S. Energy Efficiency*

Market: Generating a More Complete Picture, concluded that "our nation is not aware of the role that energy efficiency has played in satisfying our growing energy-service demands.... . The contributions of efficiency often remain invisible."

However, the tone of the ACEEE report was not one of vindication or celebration. Among the landmark news reports we have seen in our lives, few have offered such an excruciating mix of good and bad news. The good news was that America *had already found* a source for much of the additional oil and gas it needs in order to buy time for the transition to a post-oil economy—if only it would accelerate tapping into that source. "Given the right choices and investments in the many cost-effective but underutilized technologies, a variety of studies (by ACEEE and others) suggests that the United States can cost-effectively reduce energy consumption per dollar of GDP by 20–30 percent over the next 20–25 years," the report concluded. If all eight of the girders we describe are employed, the country will do even better than that.

The bad news is that if more efficient energy services are the main drivers of demand growth, as we argued earlier in this book, then greater efficiency was actually the primary cause of *increased consumption* during the past 38 years—and long before. If that is true, greater efficiency gains in the future will accelerate economic growth, and if past trends continue, more growth will be accompanied by more consumption of primary energy—which, for years to come, will mean more oil and gas. The challenge now is to break that link. It's tricky, perhaps even dangerous, to think of efficiency gains as a new source of energy, like a new oil field, although some have used that language.

One problem to be overcome is the sheer difficulty of measuring efficiency and correlating it with energy demand and growth. "In many ways, efficiency resources and investments are hard to observe, to count, and to define because they represent the energy that we *don't* use to meet our energy demands," wrote the ACEEE report's authors, Karen Erhardt and John (Skip) Laitner. "And the energy that we don't use, almost by default, becomes the energy we don't see." What we don't see, we don't measure. Another more fundamental danger is that greater efficiency often results in lower costs, and lower costs tend to drive increased demand. This phenomenon has been called "the boomerang effect," and it has been one of the excuses policymakers cite for not focusing on efficiency in the first place.

But the boomerang effect really applies to the cost of energy services or useful work, as we have called them. Cheaper services will stimulate greater consumption of those services; and across the economy, cheaper services will continue to drive economic growth, as measured by the consumption of those services. The challenge we face is obvious: Energy services must continue to get cheaper, faster, even as primary energy gets more expensive. Increased efficiency is essential to achieve this. The ACEEE report implies that past efficiency gains have not been fast *enough*. They have accounted for only three-quarters of added consumption. In the future, we need to improve efficiency fast enough to account for approximately 125 percent of service-consumption gains. We need to increase efficiency fast enough to overcome the boomerang effect and actually *reduce* the consumption of primary energy (or, at least, of carbon-based fuels). Luckily, the potential for that kind of efficiency gain is within reach.

One of the reasons for underestimating the sheer magnitude of potential efficiency gains as a resource is the widespread assumption that, although past gains in efficiency were good, the best efficiency improvements have already been made. It's as though efficiency itself is a limited resource and we are getting close to exhausting the supply. This perception isn't new. When the U.S. Congress Joint Committee on Atomic Energy was holding energy hearings in 1973, the committee requested that a report be prepared to summarize the situation for Congress and the public. The resulting report, *Understanding the National Energy Dilemma*, featured a three-dimensional "energy display system" that was impressive in its graphics but curiously lacking in scientific explanation. It was widely distributed around the country.

What was the national energy dilemma? According to the report's author, Jack Bridges, it was that demand for electricity was outrunning supply, and hundreds of new nuclear power stations would be needed. It was an argument against relying on efficiency gains. Ironically, the 1973–1974 energy crisis solved the immediate problem (much as the worsening recession knocked down the 2008 spike in gasoline prices), resulting in a sharp drop in the rate of growth of anticipated electricity demand. Most of those new nuclear power plants weren't needed or built. But the most interesting part of Bridges's argument was his calculation of energy efficiency. Without any cogent explanation, Bridges claimed that the United States was utilizing energy with an overall efficiency of nearly 50 percent. This number seemed so implausible that one of us (Robert Ayres) undertook an alternative calculation based on physical principles, and found that the real efficiency of energy use in the U.S. economy during those years (disregarding the previously mentioned problem of payload efficiency) was actually about 10 percent. The disparity between that number and Jack Bridges's figure is so large that we need to digress for a moment to explain it.

Energy efficiency is a slippery concept. On the surface, it's a simple ratio between an output and an input. That's perfectly satisfactory as long as the inputs and outputs are measuring the same thing in the same way. However, this is tricky in the case of energy because, as explained by the First Law of Thermodynamics, "energy" is actually a conserved quantity. *Conserved* means not used up. It means that the total energy flow out of any physical process or transformation is always equal to the flow of energy going into the process. No gain or loss occurs, so the transformation efficiency must be 100 percent, by definition. Therefore, the first problem with Bridges's document and the more recent Wikipedia display (and almost all public discussions of energy) is that these discussions aren't really about energy; they're about the *useful component,* which the people having the discussion don't define. However, there is a technical definition of useful energy, which is energy that can do *useful work.* The technical term is **exergy**.

The distinction between the simplistic "simple ratio" measure of energy efficiency and the technical measure of real output (exergy) is widely ignored. For example, companies advertising the (socalled) energy efficiency of gas hot water heaters or other equipment ignore this distinction. A company might claim that its heater is 85 percent efficient because 85 percent of the heat from the burner goes into the water and, therefore, only 15 percent is lost "up the stack." The arithmetic might be correct, but it's grossly misleading because *it incorrectly implies that heat at a low temperature is just as useful (in the sense of being able to do work) as heat at a high temperature.* The temperature of the gas flame in the heater is very high, about 1,800° Kelvin above absolute zero, but the temperature of the hot water that emerges is only a few degrees above room temperature (about 300° Kelvin above absolute zero).

Luckily, the exergy contents (per kilogram [kg]) of all fuels and other common materials are known and can be found in standard reference books. So the correct way to measure efficiency is to treat both the input and output in exergy terms. The input (such as natural gas) has a high exergy content because it burns at a high temperature, but the output (warm water) has a very low exergy content because it's only slightly above room temperature (compared to the flame heat). Therefore, the real exergy efficiency of a hot water heater must be very small, because the amount of heat produced by the flame could have done a lot more work than it actually ends up doing when you wash dishes or take a shower. In effect, *the heater wastes most of the temperature difference between the gas flame and the water.* In exergy terms, the efficiency of water- and space-heating systems was (and still is) only about 5 percent. In exergy terms, much the same problem applies to all of Bridges's calculations.

As we explain in an endnote for this chapter at the back of the book, the Bridges report—like the hot water heater sales departments—reached its conclusions by ignoring the Second Law of Thermodynamics. That law says that we live in a universe of irreversible exergy destruction (more commonly known as **entropy increase**): As the heat moves from flame to faucet, most of the heat exergy that was in the flame is lost. But the 1972 Bridges report and the heater brochures don't mention that. The resulting misconception has persisted not just in equipment advertising, but throughout the economy. And it appears to be one of the reasons the Bush–Cheney administration—and the media—were so disinclined to see further efficiency gains as having anything more than marginal potential. They thought the country was already doing pretty well on that front. It wasn't, and it isn't. Instead of having very little room for further improvements, we have a lot—as Robert Ayres's calculations in 1976 suggested and the ACEEE report confirmed 32 years later. Yet even after the great gains the ACEEE reported, we estimate that overall U.S. energy efficiency today is no better than about 13 percent. (By comparison, Japan is using exergy with an efficiency of more than 20 percent.)

The significance of this history—and of what we are about to show—is that American industries, institutions, and communities now have a historic opportunity to increase energy (exergy) efficiency and, simultaneously, cut energy costs and greenhouse gas (GHG) emissions on a much larger scale. That might sound suspiciously like the promise of a "free lunch." We acknowledge that, in this era of corporate fraud, identity theft, and disinformation, Americans are justifiably mindful that when it comes to a free lunch, there's no such thing. It's also true that if you start an economy from scratch, with full and honest accounting, nothing that goes into it is free (in exergy terms)—not even the water or air. But if you start with an established energy economy that is sitting on vast amounts of already-invested capital yet is operating at only 13 percent efficiency, and then you improve that performance to 20 percent or more at no net cost, the gain *as measured from where you started* is, indeed, a free lunch. And what if you run a business and you manage to make that improvement in efficiency while *increasing* your revenue? Then instead of it being a free lunch, it's a lunch that—as Amory Lovins of the Rocky Mountain Institute puts it—"they pay you to eat!" In economic terms, the improvement has a negative cost. At the corporate level, it's a very good investment—in some cases even a windfall profit. At the national level, it's economic growth without added resource depletion. That's what the United States now has a golden opportunity to achieve.

A Profitable New Pragmatism

What caused the "invisible energy" boom that the ACEEE identified? An efficiency gain from 10 to 13 percent in 36 years isn't really a boom. But even that was enough to satisfy three-fourths of new demand for consumption of energy services. That raises the question of what a real boom could do.

The reason for the gains that have been made so far can be summarized in three words: *returns on investment.* Some of those returns have been dramatic. An Oregon company, SP Newsprint, discovered in 2006 that it could reduce its energy use by increasing the amount of recycled paper it used instead of wood chips to make pulp for newsprint. The change cost the company $6.7 million but reduced the plant's energy expenses by $2.8 million a year—a payback of less than three years, with the prospect of substantially increased profits thereafter. The energy the company is now saving would be enough to provide power for nearly 5,800 Oregon houses each year.

In the same year, the University of Cincinnati installed lighting retrofits for its 6 million-square-foot campus in Ohio at a total cost of $975,793, bringing energy savings of almost 29 million kilowatt-hours and cutting the university's annual energy costs by more than $1.3 million—a payback of about 9 months. After the payback, the university's savings were enough to give full scholarships to about 25 students a year. (Given the disruptions of the subsequent economic contraction, we doubt that the savings were actually used for that purpose.) The conversion also cut 52 million pounds of carbon dioxide emissions from the Cincinnati area's contribution to global warming. And in Idaho, the J. R. Simplot Company's potato-processing plant found that, by installing new burners equipped with a different kind of controls and redesigning the system it used to provide air for combustion, it could increase the efficiency of its operations enough to save $329,000 a year in energy costs. The improvements paid for themselves in 14 months.

It doesn't take rocket science to make small changes that bring some extra profit and good public relations, as SP Newsprint, the University of Cincinnati, and the Simplot Company all did. It's another matter to challenge the basic structure of the economy, which they did not. Structural changes, which we discuss in later chapters, will add even more. But in the meantime, modest improvements across a spectrum of industries add up, not just because of what they can do for the nation's corporate profits and economic output, but because of how they can begin to change the business culture. If more Americans awaken to the realization that environmentalism is not a special interest of "tree huggers," but an essential underpinning of economic stability, the first small steps on the path to a sustainable economy might ease the way politically to the larger steps that we'll soon need to take.

Obstacles on the Trail to the Gold

The discovery that it's possible to make changes that cut energy use *and* increase profits could ignite a new gold rush in the U.S. industrial economy. However, that doesn't mean businesses won't encounter major obstacles. In the California Gold Rush of 1849, prospectors had to follow treacherous trails over the Sierras. One of those trails, which can still be trekked from present-day Squaw Valley to the town of Auburn, is a tortuous 100-mile-long wilderness path that can take an ambitious hiker up and down 2,000-foot-deep canyons and through ankle-deep snow and 110°F heat in the course of a single day. Some of the 49ers never made it. Along the way, the trail passes through the remote ghost town of Last Chance. For many American businesses, the next few years are indeed a last chance.

One danger is that even if we can prove that the "climate action will hurt the economy" mantra is wrong, and that economic survival requires investing soon and substantially in transitional and post-oil technology and infrastructure, we might have little incentive to invest in protections that won't visibly pay off in the next 10–20 years. To the average American on a day when no tornado or hurricane is on the horizon, the world might look quite normal and stable. Because ExxonMobil, Shell, and BP have done a slick job of telling us that there's an "ocean of oil" still under our feet, not every business manager will be convinced that installing energy-saving modifications at the plant matters that much. Some recall that the panicky response to high oil prices in the 1970s was followed by $10-per-barrel oil in the 1980s, and that the same kind of recovery happened—even faster—in late 2008. A study of executive attitudes reported by The Alliance to Save Energy found that many executives believe energy-saving programs are just "technical" matters best left to the engineers, and are not significant aspects of business strategy.[2]

A related reason we can expect resistance to any significant change of corporate direction is that, even when people do see future

---

[2] The report, *Executive Reactions to Energy Efficiency*, noted: "While executives are keenly aware of energy prices, they are usually unconcerned with energy consumption. This is why energy efficiency is often delegated to plant engineering and maintenance professionals. It is expected that *technical* people will devise *technical* solutions within the confines of their areas of authority. Engineers naturally focus on technology and hardware, while operations or production managers oversee the daily behaviors and procedures that directly impact energy consumption. Procurement directors often make equipment purchases based on the immediate cost of acquisition, not total operating costs. Unfortunately, these different interpretations of energy cost control are reinforced by departmental "turf" issues, which preclude the kind of cooperation that is necessary for an organization to become energy efficient."

consequences, personal attachment—to power, control, or sense of self-worth—can trump an already tenuous sense of intergenerational responsibility. A 60-year-old CEO might have 20 more years to live. If he has a large surplus of personal wealth, he can afford to insulate himself and his family from the early ravages of climate change. He can acquire a gated estate on high ground for his retirement, hire security guards, and obtain ample supplies of gasoline even if it eventually reaches $20 a gallon, and ample supplies of beef even if it reaches $100 a pound. If he doesn't want to do anything for those who live a decade or two after he's gone, he doesn't need to. Some CEOs *don't* seem to care. And for them, the efficiency gained by such routine modifications as upgrading the lighting in their stores or factories isn't even a blip on their radar.

Some of the resistance may result from neither ideological rigidity nor personal weakness, but from misunderstandings by business managers about what efficiency really is or how they can achieve it. For example, the aforementioned study of executive reactions notes that many corporate leaders genuinely believe that their facilities are already efficient. Part of the misunderstanding arises from what the Alliance to Save Energy calls an "obsolete paradigm for making energy decisions." Many of today's industrial organizations "continue to manage energy use and costs in the same way that they did in the 1980s—an era of cheap fuel and regulated utilities." The old paradigm was characterized by three perceptions:

1. "Energy is not our core business." Energy was given the marginal attention and budgetary resources commensurate with its status as a support (or secondary) function instead of a controllable cost of production and a source of recoverable earnings.

2. "Low fuel price is the solution." The job was handed off to the procurement director or, indirectly, to the government, to get fuel at the lowest price possible.

3. "You're the engineer; you figure it out!"

These misunderstandings might help to explain why, after 30 years, still only a minority of U.S. businesses have fully tapped into the efficiency reserve. To put that in perspective, consider that of the approximately $300 billion invested in energy efficiency in the United States during the 38-year span the ACEEE studied, a larger share

(29%) went to consumer electronics and appliances than to industrial operations (25%). U.S. factories and refineries have a long way to go.˙

The resistance usually involves some combination of these factors: An executive might be well aware that energy efficiency is a growing factor in business performance, but believe that his company is already cutting-edge, or that his crackerjack engineers will take care of any energy glitches. He might care about his kids' future, but with them now out in the world on their own, he might not be inclined to make any large sacrifice of the pleasures and perks he has won—including the consummate pleasure of continuing to win at the capitalist game. And he might have unquestioning belief—because of what the prevailing economic doctrine has taught him—that his kids will inevitably be even richer than he is, and that they'll be beneficiaries of even greater technological progress and economic growth for as long as they live. No matter how compelling the evidence of human-caused warming driving climate change becomes, and how compelling our refutation of the dominant economic doctrine might be, many CEOs and others who have their hands on the wheel of our ship will refuse to venture a radical change of course. ExxonMobil, Edison International, the "Clean Coal" lobby, and the people who think Earth was created 6,000 years ago with the coal already made, are among them. In building our bridge to the post-oil future, we're in for a fight.

However, the experiences of businesses such as SP Newsprint and J.R. Simplot suggest a strategy that can shorten the fight. Mainstream economists still cling to the view that examples like these are anomalous exceptions because their theory says that if opportunities for added profits really existed in any significant numbers, enterprising entrepreneurs in our supposedly free competitive-market system would immediately exploit them. So the free lunch can't really be there. The "low-hanging fruit" would be gone.[3] If we can show that such opportunities are *not* exceptions, but are virtually omnipresent and do not have quickly diminishing returns, we have a means of

---

[3] Critics of this view tell a joke about a man walking down the sidewalk with a friend who is a neoclassical economist. The man sees a $100 bill lying on the ground and leans down to pick it up, but the economist says, "Don't bother—if it was real, someone would already have taken it."

buying valuable time in the race against peak oil and climate change. If we can show the wider public that businesses everywhere can make changes that reduce energy use and increase profits within a year or two, even some of the most intractable resistors will be attracted. Quick returns on investment should draw relatively few objections even from those who favor their own comforts over those of the next generation. And for people who remain wedded to the doctrine that future generations can take care of themselves better than we can, there should be no objection to actions that pay off almost immediately—for the present generation as well as the next. If a small minority of U.S. businesses could achieve a de facto doubling of economic output per barrel of oil or ton of coal in the past 30 years (a halving of energy intensity as reported by the ACEEE), imagine what could be achieved if tens of thousands of the holdouts joined in.

The Myth of the Disappearing Fruit

The economists' argument about the quick disappearance of low-hanging fruit might be a persuasive metaphor, but it doesn't hold up in the modern industrial economy. An orchard stands fairly still, but most industrial operations are moving targets, continuously introducing new processes, practices, and equipment. Processing plants, refineries, and factories are places of enormously dynamic complexity. One eco-fix might actually create another opportunity to save—increasingly with the aid of advanced sensors, automated control systems, or other energy-tracking tools.

Sometimes the new crop of low-hanging fruit comes from a technical development of the sort that the traditional economists describe as exogenous. For example, successive generations of computers have become exponentially less energy-consuming in recent years. In 1996, the first Intel supercomputer capable of a trillion calculations per second consumed 500,000 watts (W). By spring 2007, the company had made a dime-sized chip of equivalent power that consumed just 62W. In the context of our earlier discussion of economic growth theory, this "technological progress" wasn't just an example of dramatic progress in miniaturization or operational speed, but in *making energy cheaper* for a particular energy service.

As experience demonstrates, a company that makes a concerted effort to find and exploit such new opportunities keeps finding them.

This reality has been documented since the 1970s, when the emerging environmental movement—and the mandates of the new Clean Water and Clean Air acts—put American industries under pressure to reduce the huge quantities of toxic wastes being dumped into the commons. Companies needed to document any progress they made in the cleanup. As a result, we now have hundreds of success stories. The motive for this record keeping was not so much the problem of energy consumption, as of pollution (the oil crises of the 1970s were quickly forgotten when gas prices dropped back), but energy efficiency and waste reduction were closely linked. Moreover, the earlier in the chain of processes the waste could be stopped, the more successful the effort would be, because it's cheaper to prevent pollution than to try to catch it at the smokestack or—worse—try to clean it up after it has been dispersed into the air, water, or ground. A number of companies instituted programs to "prevent pollution." And as their experience proves, measures that prevented pollution at the outset did indeed reduce energy use. These programs demonstrated a fundamental principle of industrial ecology: An operation that mimics nature by recycling its waste—including its waste-energy streams—puts less waste into the environment.

One of the first large companies to adopt this principle was the Minnesota-based 3M Corporation (formerly Minnesota Mining and Manufacturing), which started its 3P program (Pollution Prevention Pays) in 1975. During the following 20 years, it implemented 5,600 in-house projects that reduced manufacturing waste by a cumulative total of more than 1 million tons and earned the company more than $1 billion in *first-year* returns on investment (ongoing returns were not measured).

In the 1980s, Atlanta-based Interface Inc., the world's largest manufacturer of modular carpet, began a campaign to become one of the first environmentally sustainable companies. In 1994, Interface founder Ray Anderson had what he described as "an epiphany that the current business system was wreaking ecological havoc on the planet." Between 1996 and 2006, his company cut its solid waste by

63 percent and its greenhouse gas emissions by 46 percent; in doing so, it reduced its energy use per unit of product by 28 percent.

About the same time, in Colorado, Coors Brewing Company switched to UV-imprinted coatings for its beer cans, eliminating the use of solvents that emitted volatile organic compounds (VOCs). The switch greatly reduced the company's output of hazardous waste. It also greatly reduced energy use, because Coors no longer needed to fire up large ovens to cure its cans.

Other large companies as well have reduced energy use by reducing pollution. Fedex, in collaboration with the nonprofit group Environmental Defense, undertook a campaign in 2000 to reduce the emissions of its fleet by turning to specially designed hybrid trucks. By 2006, about 75 of the new trucks were in service, each reducing emissions of soot by 96 percent and nitrogen oxides by 65 percent—and improving fuel efficiency by 57 percent. The company's goal is to convert its entire fleet of 30,000 trucks as soon as possible.

Many hundreds of such cases have been documented, and many have paid off quickly. *There could be tens of thousands more.* While the neoclassical economists may find it hard to believe, many of the programs have been initiated by government, whether via the carrot of incentives or the stick of regulations. Once initiated, the programs have consistently increased profits. For example, the U.S. Department of Energy (DOE) reported in 2008 that its "Save Energy Now" campaign had assessed 543 industrial plants where DOE technicians had worked with the plants to make technical modifications, and these modifications had produced aggregate energy cost savings of $706 million during the previous three years. The modifications cut the plants' use of natural gas by enough to power 1 million houses for a year, and reduced carbon dioxide emissions equivalent to the exhaust from 1.2 million cars.

The New York State Energy Research and Development Authority (NYSERDA), a public benefit corporation launched in 1975, reported in 2008 that, since 1990, its research and development program had successfully brought into use more than 80 energy-saving products or processes. For example, at the Ames Goldsmith Company in Glens Falls, New York, which supplies 96 percent of the silver oxide and 12 percent of the total silver used in the United States, a new process was developed to recover silver from spent alumina catalyst. The innovation reduced the company's water and energy consumption by 33 percent, saving it $130,000 a year while increasing its output capacity by 50 percent. In another case, Gould's Pumps of Seneca Falls, New York, was concerned about its VOC outputs, which are highly toxic to employees and the public. For 35 years, Gould's had been coating its industrial pumps with a solvent-based coating (similar to the one Coors had been using for its beer cans) containing more than 5 pounds of VOCs per gallon. With the help of NYSERDA, Gould's joined with Strathmore Products, Inc., of Syracuse, to develop a water-based coating that reduced VOC content to less than 1.7 pounds per gallon. It then worked with Optimum Air of Malta, New York, to develop a system to quickly dry the surface. The change saved Gould's $183,000 in annual energy use while adding $800,000 to its sales due to improved product quality. By 2008, NYSERDA had provided energy-saving, revenue-boosting assistance to more than 400 companies—a small fraction of those in the state, but enough to make the case that such opportunities abound. In its landmark 2008 report on the U.S. efficiency boom, the ACEEE reported that, in 2004 alone, efficiency technologies saved $178 billion in the buildings sector, $75 billion in the industrial sector, and $33 billion in the transportation sector.

The Myth of Diminishing Returns

American businesses have been handicapped not only by the widely held belief that energy efficiency is only a marginal resource, but by the corollary belief that even when businesses can find such efficiency gains, they're "one shots," and that investing in these gains won't pay in the long run because of inevitably diminishing returns. In theory, it's true that the rational strategy is to take the best first, whether from oil fields or gold mines. But the flaw in the theory is that it's rarely possible to know in advance which opportunities will be the most profitable. In reality, the first choices are likely to be the easiest and most visible.

We found a dramatic illustration of this point in a program the Dow Chemical Corporation instituted three decades ago. In 1980, an engineer named Kenneth Nelson, who worked at Dow's Louisiana

Division (a plant located in the so-called "Cancer Alley" area of the state), proposed to his management that it launch an in-house contest to find ways to reduce the plant's chemical wastes by improving the efficiency of its manufacturing processes. Dow management approved, agreeing to fund any projects costing $200,000 or less that could pay back that cost within a year. In the first year, the company's midlevel engineers proposed 39 projects, and 27 of them were funded at a total cost of $1.7 million. At the end of the year, the projects had produced a cumulative return on investment (ROI) of 173 percent—a payback of about seven months.

The contest continued, and contrary to what the economists in Washington would have predicted, the returns did not diminish in the second year; they substantially *increased.* Dow named the program WRAP (Waste Reduction Always Pays), and it lived up to its name for the next decade and beyond. In the tenth year, the engineers added 108 new projects that yielded an average one-year ROI of 309 percent. In the eleventh year, they added 109 projects with an ROI of 305 percent. And in the twelfth year, there were 140 projects—the most yet—with an average ROI of 298 percent. The twelfth year brought more profit to Dow than any of the previous 11. Yet after that twelfth year, the WRAP program—which had brought Dow more than a billion dollars in additional profits—was abruptly discontinued. *Why?*

One answer is that the system of risks and rewards for top executives encourages corporate growth, not efficiency. It doesn't work that way in every company, because Americans are divided and some companies have now begun to genuinely embrace their environmental and intergenerational responsibilities while others have not. But in the prevailing corporate system, opportunities to increase gross revenues too often trump opportunities for greater efficiency and profitability. When a division grows, its manager is promoted, gets a larger office and a bigger salary, and has more people working for him or her. An engineer who saves money is thanked and might get a small bonus, but not much more. Shortly after the Louisiana plant shut down its WRAP program, Nelson left.

However, subsequent assessments of what happened in its Louisiana contest may have begun to change the corporate culture at Dow. Between 1995 and 2005, energy conservation was made a central component of the company's business strategy. According to a Dow spokesperson, during that period the company achieved an additional savings of more than $5 billion, with comparable returns on investment. The Dow experience leaves us with an important affirmation: The opportunities to improve energy efficiencies at a profit do not quickly disappear. Well into the twenty-first century, our industrial and consumer economies are still riddled with energy-wasting products, processes, and practices. We hear plenty about the consumer and building-sector examples (gas-hogging SUVs, imported foods, incandescent light bulbs, and poorly insulated houses) but almost nothing about the ones behind industrial gates. Now, with the ACEEE's dramatic evidence that *energy efficiency can provide vastly more muscle to the U.S. economy than new supply*, that might change.

The Dow experience also suggests that if shareholders knew more about how often executives have sacrificed profits for sales or market share (a game that has high rewards for CEOs but diminishing or even negative rewards for investors), they might have second thoughts about their votes for corporate directors, not to mention bonuses. A great irony arises here. Left-wing criticism of American capitalism has long focused on the presumed obsession with profits (at the expense of social well-being) as a root of America's rich–poor divide and environmental destruction, among other failings. But in the game many top executives play, winning is not measured by the efficiency and profitability with which they produce their services or products, but by the power they wield in their markets.

Historically, Americans have abhorred monopolies as killers of competition (and producers of robber-baron levels of profitability), resulting in the antitrust laws that have been so long on the books but that we hear little of now. In recent decades, the news has been more about a particular path toward monopoly or oligopoly—the intoxicating path of mergers and acquisitions. Yet those deals aren't generally conducive to greater profitability. The quickest way to build market share is to buy or hijack your competitors and add their market share to yours. In the game of personal power, people who bring mergers get huge rewards—or, at least, that's what was happening before the whole economy hit a wall. But investors do not get these huge rewards. In a recent study of 700 corporate mergers, the international accounting firm KPMG found that only 17 percent of the deals created real value for the shareholders, and more than half *destroyed* value. More important, society at large doesn't benefit, because as concentration of an industry increases, competition weakens—and so does the incentive to use energy more efficiently.

A Sea-Change in Corporate Culture

The double-dividends strategy still leaves the question of how to bring the resistors around to the all-important decision to invest in larger, *long-term* changes. One answer is that energy efficiency buys (a little) time to change the demographic and political equation. It's an essential girder of the energy-transition bridge. During the next few years, some of the more myopic CEOs will be replaced by people 10 or 20 years younger. Statistically, the outlook will be somewhat different for the new managers: Their life expectancies, instead of extending 20 years into the climate-change era, will extend much longer—into a time when the gated compound and capability to buy what the rest of us can't might not be so protective. Economics will also see some changing of the guard. The economic graduate students of today will (hopefully) not be as dismissive of science as were their predecessors of the 1980s, when the physical sciences were still associated in the public mind with the era of atomic bomb building, hazardous chemicals, and space exploration, and the IPCC didn't exist. We also expect that, within the next few years, the story of the cost-share theorem will wind its way through the economic journals and into the classrooms of Economics 101. Economists' understanding of how the price of energy drives economic growth should bring a change in their teaching and in the advice they now feel free to give to business and government. That, in turn, will enable business managers to see energy efficiency as far more central to their success.

Not all top corporate executives are dreaming of a return to the good old days of guaranteed growth; at least a few have taken the wider implications of their work to heart. At General Electric, Chairman Jeffrey Immelt announced in 2005 that his company would reduce its greenhouse gas (GHG) emissions by 1 percent by 2012—which would require making the company's emissions 40 percent lower than if it took no action, while increasing sales as projected. Maybe Immelt was being conservatively coy (some accused GE of greenwashing, and Immelt knew that he'd be excoriated by people like us if he didn't come through). But by the end of 2006, less than a year after the campaign began, GE was able to make the surprising announcement that not only that its sales had run much higher than publicly projected, but that, despite those increased sales, its GHG emissions had declined by 4 percent—far better than Immelt had promised for 2012, and six years ahead of schedule. And executives such as Immelt, or Interface's Ray Anderson, are now far from alone. Since 1992, an alliance of major U.S. and global corporations, the World Business Council for Sustainable Development (WBCSD), has brought together a core group of like-minded executives. The WBCSD is only a tiny minority of the tens of thousands of companies that will need to discover the rewards of double dividends if the bridge is to work. But at least from now on, any board of directors or CEO hoping to move in that direction won't need to bushwhack through uncharted territory. There's now a well-blazed path to the gold.

5

The Future of Electric Power

During every economic setback since the 1930s, whenever any of us worried that we might be headed for another Great Depression, economists assured us that it couldn't happen again because "we now have safeguards they didn't have back then." And as the decades went by, history seemed to confirm that sanguine outlook. But in the last two years of the second Bush administration, those alleged safeguards failed. (As with the Emperor's new clothes, they didn't really exist.) Self-regulation didn't work. The Federal Reserve miscalculated, Wall Street traders ran Ponzi schemes, the SEC looked the other way, and the FDIC didn't have enough cash to back up all the failing banks.

After the oil crises and disillusionment of the 1970s, some of us worried about whether something like *that* might happen again— only with more lasting impact, because by then we knew that oil would become increasingly scarce during our lifetimes. The first warnings of peak oil began circulating after M. King Hubbert's shocking (and accurate) prediction in 1957 that U.S. oil production would peak in 1969–1970. Experts assured us that it wasn't true and that safeguards had been established. But Hubbert was right, and the safeguards were illusory.

In 1978, In the wake of the 1973–1974 "oil shocks," Congress passed a sweeping new law, the Public Utilities Regulatory Policy Act (PURPA). It was intended to increase electricity output without needing to build a lot of new power plants. Where monopolies had operated, competition would arise; where the United States had been at the mercy of Middle-Eastern oil dictators, it would build vibrant new industries of solar and wind power. "Energy independence" became a clarion call.

However, after oil and gas prices fell back, public worries abated and the new competitive electricity-supply system that PURPA envisioned was never realized. Three decades later, as oil and gasoline prices spiked again in 2008, and with half of the voters too young to remember the oil embargo of 1973–1974, every public official boldly called for energy independence as if it had suddenly come to each of them in an epiphany. Once again, oil prices soon fell back, but this time the call for independence was not so easily forgotten. We were now three decades closer to global peak-oil output; 2 billion more people lived on the planet than in the 1970s (hundreds of millions of them hoping to drive cars and have air-conditioning); and six years of Iraq War had made it clear that military defense of access to oil—if that's what it was[1]—is unaffordable.

What happened to PURPA? The short answer is that nobody with any clout was in charge. It failed in its mission as badly as those in charge of the Federal Reserve, Fannie Mae, Countrywide Bank, and Wall Street failed in theirs. It didn't *entirely* fail, because in one limited but important way, it liberated a new energy source. In Chapter 2, "Recapturing Lost Energy," we noted that about a thousand U.S. companies are producing electricity by harnessing waste-energy streams (heat and steam), and a revision of the PURPA passed in 1992 enables those companies to sell any excess electricity they produce back to the grid at prices the state utility commissions set . But as we noted, these companies account for only 10 percent of the national waste-stream potential. Most of the other 90 percent of the potential has been blocked, mainly because PURPA didn't allow companies to sell electricity directly to other companies. And on a larger scale (apart from sellback of excess power by other industries such as the steel plants we described earlier), the government has blocked real competition in the utility industry itself.

PURPA has failed in its principal mission for two reasons. First, it limits the prices new competitors can get to the so-called "avoided costs" of the established utilities—meaning that they can only be paid

---

[1] After retiring as chairman of the Federal Reserve, Alan Greenspan wrote in his memoirs, "I am saddened that it is politically inconvenient to acknowledge what everyone knows: the Iraq War is largely about oil."

a price that is less than what the utility would have to pay to produce that power. Utility accountants and utility-friendly state commissioners fix that price, so independent producers—*especially* producers of wind or solar power—have usually found it impossible to compete. In the pricing, the producers of renewables get no credit for producing emissions-free power, or for reducing the vulnerability of the grid to the loss of an important link due to accident or terrorism.

That barrier has remained so obscure—so rarely mentioned in public discussion—that when a man named Bill Keith brought it up in a nationally televised town-hall meeting with President Obama in February 2009, media reports missed it entirely. The meeting was held in economically devastated Elkhart, Indiana, and the President answered questions from eight members of the audience. When Keith's turn came, he drew applause when he introduced himself, saying, "I manufacture a solar-powered attic fan right here in Indiana." He then turned from his own business to a more general concern: "What we need is a more friendly environment from the utility companies, so if I want to put a solar system on my house, I can get more than 9¢ on the dollar for the electricity I feed back into it... . Those of you out there that think the prior administration or someone gave us some kind of benefits for being a green company here—there are none."

Obama responded with enthusiasm but appeared to miss Keith's key point: Federally mandated renewable-energy output from the utilities won't help new businesses ramp up the country's renewable energy production, as long as the electric utilities can continue their price-fixing. In the days after the town-hall meeting, environmental and green-business blogs gave much attention to the exchange, but their focus was on the great publicity given to Keith's successful, made-in-America solar business, not on the lack of incentives for renewable-energy production. They, too, missed the key point.

PURPA's second failure is that although it "encourages" alternative energy, the federal law has no teeth. PURPA defers to the states to implement the mandate, and not all states have taken an interest. Louisiana, a state where the oil and gas industry is king (Louisiana is the number-one producer of oil and the number-two producer of natural gas among the 50 states), has simply ignored PURPA. South Carolina, Kentucky, and South Dakota have also ignored it. In the rest of the states, it's used only as a handy means of postponing costly capital investments in new plants, while prohibiting any real competition from plants offering clean energy from renewable sources.

One of the major girders of the energy-transition bridge must be a genuine restructuring of the electric utility industry—a reform that achieves the original purpose of PURPA without being hijacked by the industry being restructured. This restructuring might be the most politically difficult step in the whole bridge-building enterprise. The utility industry is not only one of the largest sectors of the U.S. economy, but also one of the most opaque and secretive.

This girder alone offers potential economic savings comparable to the cost of the $700 billion financial sector bailout of 2008—the one that, based on what we now know about the "third driver" of economic growth, *has to be paid for by a reduction of energy service costs.* As we explain shortly, the outfall from utility reform will ultimately yield hundreds of billions of dollars of savings, while cutting global warming emissions by a percentage that most economists have said could never be accomplished without "hurting the economy."

We hasten to add that *in the later stages* of the restructuring, the utility sector shouldn't need to be dragged kicking and screaming, although its initial resistance will be fierce. To draw a quick analogy, IBM Corporation, once the goliath of mainframe computers, has adjusted quite well to the omnipresence of minicomputers and personal computers that have largely replaced those bulky mainframes. However, a core question for a nation building an energy bridge is how long it will take for the regulators and managers of the utilities to recognize what their future will be and how cooperative they need to be in facilitating—and not obstructing—the transformation.

We can begin the recognition process by drawing a picture of what the mid-twenty-first-century electric power system will look like if all goes well. In the next few decades, a plethora of independent, local sources of power will join the grid, which now serves almost every light bulb in the land by delivering electricity primarily from large, centralized power plants. Today, electric generation and distribution to the grid is a monolithic system radiating out from those giant coal- or natural gas–burning (or nuclear or hydroelectric) plants, with local users having no control of either the security of their power supply or the price they pay. (Consider the growing incidence of power failures and brownouts in recent years, and the growing difficulty of understanding or questioning your electric bill.)

In the future, the distribution system will have a new mandate: not just to deliver power as it does now, but to provide (1) a more coordinated and efficient integration among multiple suppliers and consumers of different sizes and needs, including differences in peak-load times, and (2) a more secure and reliable protection against power failures and leaks.

The micropower revolution won't displace the grid, but it will greatly improve its efficiency. No longer will an individual home and a giant aluminum smelter necessarily get their power from the same cumbersome system—an elephant and a mouse drinking from the same dish. Central plants will provide backup power to everybody, but they'll focus on serving clusters of large users such as factories, pumping stations, and water desalination plants (see Chapter 9, "The Water-Energy Connection"), and on sending power from regions that have excess to those that don't have enough.

The high-voltage transmission lines will be needed not only to continue delivering power to cities from distant sources that can't be moved, such as hydroelectric plants or wind farms, but also to take advantage of peak-load differences between time zones. For example, if power is being generated in California at 5 AM, when real-time demand is low, utility companies could transmit the excess at that hour to the Midwest or the East during their much higher—and simultaneous—7 AM or 8 AM rush-hour demands. And as wind and solar power come increasingly into the picture during the transitional years, the grid will be able to transmit electricity from places where the wind is blowing or the sun is shining to places where it's not. As President Obama acknowledged in early 2009, "If we're going to be serious about renewable energy, I want to be able to get wind power from North Dakota to population centers, like Chicago."

At the same time, local power will be generated more locally. In places of high building density (such as central cities, small industrial complexes, business parks, shopping centers, airports, hospitals, universities, and military bases), the new system will provide *two* services—both power *and* heat (CHP). It will need to be an open, two-way system, enabling customers to sell any excess power or heat they produce either back to the grid or directly to neighbors, bypassing the grid. Since PURPA was passed and revised, the government has allowed selling back to the grid in some places, but not selling directly to other users across the street.

Generating units should be increasingly small—initially, furnace-sized CHP units inside individual buildings or in conjunction with adjoining facilities, as we briefly discussed in Chapter 2, but eventually extended to an open, free market for all producers and users. As the big, obsolescent coal-burning power plants depreciate, we can replace them with clusters of smaller, decentralized units that also sell heat, and dismantle many of the older (and very ugly) transmission lines in and around the city.

The "Coal-Rush" Distraction

One impediment to the decentralization of electric power is the political power of the coal industry and its dependents. Coal is the largest fuel source for electric power in the United States. The country has more than 600 coal-fired power plants, accounting for 48 percent of the energy electric utilities consume. It's the one source that will be shut out when power generation moves back to local neighborhoods. Long ago, many American homes had coal chutes into their basements. That day is gone. But coal continues to fire the central plants, which are located in places where the smoke and sulfur dioxide were less noticed for much of the mid–twentieth century—until environmental scientists began noticing the morbid effects of acid rain on forests and lakes in the 1970s, and climate scientists began revealing coal's contributions to global warming in the 1990s.

One result of those revelations has been an explosion of regional battles over plans to build new coal-fired plants—quickly, before new restrictions take effect. And as the regional battles have escalated into a national one (albeit still not much of an issue in mainstream media), the disconcerting effect has been to trigger an even more panicked firefight by the threatened coal interests—a phenomenon widely described on the Internet as the "Coal Rush."

Part of the acrimony over coal traces back to a legal loophole that for years exempted carbon dioxide emissions from being classified as pollutants under the Clean Air Act. Congress created the loophole after lobbyists argued that carbon dioxide is a natural part of Earth's carbon cycle, and that it's what plants and trees need in order to live. How could something that is essential to the growth of redwood trees and roses be regarded as a pollutant? Environmentalists were not impressed with this disingenuousness, pointing out that coal is the largest human-generated source of greenhouse gases and that the other pollutants in coal smoke, such as mercury, pose threats to both human health and ecological stability worldwide.[2]

In 2007, the U.S. Supreme Court issued a landmark ruling: The U.S. Environmental Protection Agency (EPA) has the authority to regulate carbon dioxide as a pollutant after all. The EPA, which was being tightly muzzled by the second Bush administration at that time, insisted that it had authority to regulate only mobile sources of carbon dioxide (such as cars and trucks), and that stationary power plants were therefore beyond its jurisdiction. Four months later, the EPA gave the go-ahead for a new coal-fired facility to expand the Bonanza power plant in Uintah County, Utah. The Sierra Club, the Environmental Defense Fund, and the regional nonprofit Western Resource Advocates sued on the grounds that the permit "fails to require any controls for the millions of tons of carbon dioxide that this plant would emit each year." In November 2008, the Court of Appeals ruled against the EPA and the new facility.

Around the same time, a similar battle was being waged over a proposed new expansion at the Duke Power Company's Cliffside plant in Rutherford County, North Carolina, not far from the Smoky Mountain National Park. Analysts estimated that the output of the new coal-fired plant would be equivalent to the exhaust of a million cars. In keeping with the usual cozy relationships that have prevailed between public utility commissions and utilities across the country, the North Carolina utilities commission quickly approved the project. The Sierra Club again led a suit to halt the project—and won.

---

[2] Beyond its contribution to climate change, the combustion of coal accounts for two-thirds of all U.S. emissions of sulfur dioxide—a principal cause of asthma attacks in humans and of acid-rain destruction of lakes and forests.

However, the coal-electric industry's reaction wasn't to smell the fresh air of the future and cut back its expansion plans, but instead to accelerate them. Realizing that Congress seemed very close to establishing carbon limits for U.S. industries under the new Obama administration, the industry began a rush to construct new coal plants, on the theory that they could be "grandfathered"—exempted from forthcoming carbon caps on the grounds that they had been started before the new limits were enacted. According to tracking reports, plans for more than 100 new coal-fired plants were in the works as of late 2008. The result was a frenzied tug-of-war involving utilities, commissioners, environmental organizations, federal agencies, and courts. In the turmoil, it seemed that even court decisions couldn't be relied on—much less the commitments of key players. Duke Power chairman Jim Rogers had been known to support climate-change mitigation in principle,[3] but he continued to aggressively push his new coal-fired plant.

The absurdity of the Coal Rush, from a public-interest perspective, is that it is a panic reaction—or myopic greed—of what bloggers began dubbing "fossil fools." Some rich profits could be made by pushing the human prospect a few fateful strides closer to the cliff of climatic and ecological catastrophe. But for U.S. energy independence, a plethora of new coal plants at this point would be too late, too costly, and too dangerous to human health and climatic stability. The absurdity was underscored by a letter that James Hansen, the renowned NASA scientist who had first warned the U.S. Congress of impending climate change in the 1980s, wrote to Duke's Chairman Rogers. The proposed new plant, Hansen wrote, would be "a terrible, foreseeable waste of money" and "will have to be shut down."[4]

From our perspective, the absurdity is best encapsulated by two numbers. The addition to the Bonanza facility, where the coal rush was triggered, was designed to generate a mere 86 megawatts (MW).

---

[3] That same year, Rogers told *Electric Utility Week*, "By creating a policy that places energy efficiency on economic parity with other forms of power supply, utilities will be able to meet consumers' needs through saving watts, as well as making watts, without negative financial consequences."

[4] Richard Heinberg's web site essay, "The Great Coal Rush (and Why It Will Fail)," provides an interesting discussion of this eventuality.

The Cokenergy recycling facility in Indiana, which we described in Chapter 2, produces 90MW from fossil fuel waste, completely carbon free. More of the latter, not the former, will constitute the decentralized-utility girder.

Local Power, National Security

Even when the coal troglodytes retreat (and as people breathe easier and the endangered forests get a reprieve), the big-utility holdouts and their lobbyists will reflexively trot out their old argument that central power plants are natural monopolies with economies of scale that local production can't compete with. Although central plants were indeed more efficient and economical when the centralized utility system was first established in the 1920s, they have not significantly improved their energy efficiency in the last half-century. In the meantime, the rest of the world has moved on. Technologically and economically, the industry is being overtaken by the capabilities of the new local-power, small-unit systems such as gas turbines, diesel engines, rooftop PV, small wind turbines, and even high-temperature fuel cells. The first four are available off-the-shelf right now, although rooftop PV is not yet economically competitive in most cases. Fuel-cell development is proceeding rapidly. (For more discussion of micropower systems, see Chapter 8, "Preparing Cities for the Perfect Storm"). The advantages of decentralization are now as great as the advantages of centralization were in the 1920s.

Reduced Cost

Given the impact of energy costs on our capacity to respond to *any* of the now-looming challenges to our industrial civilization, reduced cost will be the ultimate prize. Although the prize is great, it has rarely been discussed in public, for the simple reason that, for the past half-century, an electric-power monopoly has existed in every community of the United States, guarded by law and perpetuated by a rarely challenged dogma that says central power plants are inherently more energy-efficient than smaller-scale production could be. This is where it gets tricky, which is one reason the public (and Congress) has trouble keeping track. If we just count the cost of electricity generation at the plant, and if low-temperature heat has no value, the cost of central power is indeed cheaper than any competing system. Stop the discussion right there, and it's enough to keep the central-power cowboys in the saddle.

But wait a moment. Low-temperature heat obviously does have value, because a considerable fraction of the U.S. energy bill is spent every year on producing low-temperature heat for buildings. The problem is that so-called "central" power plants are too big and too far away from potential customers for that heat. So they discard it into the air or, in some cases, into lakes and rivers, where it might cause ecological damage.

The power companies report only to public-utility commissions, which are theoretically supposed to minimize electric power costs to consumers. Heat is not part of their mandate. Electricity costs to users are determined by current fuel costs plus a component based on capital costs of the steam-electric plants that actually *generate* power, plus the costs of *transmission* (those power-line rights-of-way, towers, and wires) and *distribution* (step-down transformers connecting to neighborhood wires). There is also something called *redundancy* (backup reserve), which most consumers probably don't think much about, but could have large implications for energy security in the years ahead. When those transmission-and-distribution (T&D) costs are added to the central-plant costs, and then the total is bumped up enough to meet the redundancy requirement for extra capacity at times of peak loads in the region, the overall cost is just about doubled. That makes it higher than the cost of local production, which, by definition, has less T&D and requires less redundancy. (When a greater share of the power is produced locally, an area-wide power loss is less likely. If a 500,000-volt line goes down in an upstate New York ice storm or a California wildfire, thousands of homes and businesses might be blacked out, but the same storm or fire won't black out a building that has its own power.)

The other great cost advantage of local generation is the potential for selling the not-so-waste heat associated with the system, whether it be a gas turbine, a diesel engine, or a fuel cell. Cogeneration, in which the same system produces both electric power and useful heat (usually for hot water and space heating), enables the value of the heat to be realized. It also saves energy and the pollutant emissions that accompany fossil-fuel combustion.

Lower Energy Consumption

We have already mentioned the not-very-proud fact that the U.S. electric-utility industry delivers just 1 unit of energy's worth of electricity to a consumer for every 3 units of fuel it burns—an anemic 33 percent efficiency. In contrast, cogeneration plants that a company is allowed to install for its own use, but not for purposes of competing with a utility, can operate at 50–80 percent efficiency by using the otherwise-waste heat *as heat*. Consider this fairly conservative estimate of the potential for reduced fuel use nationwide: If we captured just half of the energy that central power plants currently lose and used it to heat buildings, that heat could replace low-temperature heat that boilers currently produce by burning about 13 quadrillion btus ("quads") of fuel—*15 percent of all fossil fuel burned in the country*, and a much larger fraction of the coal that is burned. If we were a little more ambitious and eventually used two-thirds of the wasted heat, the share of fossil fuel burned in the United States would decrease by 20 percent. (It wouldn't be realistic to try to recapture 100 percent of the waste heat because some of the large fossil-fueled power plants will remain in operation for decades, even under the best of circumstances.) The 15–20 percent reduction in national fossil fuel use (mostly natural gas and coal) makes a telling comparison with the share of U.S. fossil fuel that imported oil currently provides.

Combined heat and power (CHP) is applicable only in places that can use low-temperature heat. The International Energy Agency (IEA) provides more restrictive criteria for success:

- A ratio of electricity to fuel costs of at least 2.5:1
- Heating demand of at least 5,000 hours per year
- Grid connection at reasonable prices
- Availability of space for equipment
- Short distances for heat transport

Not long ago, the IEA—with substantial inputs from utilities—estimated CHP's U.S. potential for emissions reduction to be about 4 percent by 2015. But that estimate, perhaps because of the utilities' input, clearly didn't envision any fundamental changes in the laws that restrict utility competition. We believe that the real potential is much greater. If Denmark, the Netherlands, and Finland have already achieved 40–50 percent penetration by CHP, the United States should be able to do better than the current 4 percent potential that the IEA has acknowledged.

Reduced Carbon Emissions

A strategy that reduces U.S. fossil fuel use by 15–20 percent will also reduce carbon emissions by about that same amount, depending on the mix of coal, natural gas, hydroelectric power, and nuclear power from utilities being replaced by the mainly gas-fired local power. (The emissions will continue falling as local production shifts increasingly from natural gas to solar, wind, or solar-derived hydrogen fuel cells during the coming years.) It's instructive to compare that 15–20 percent cut in greenhouse gas emissions with the 3–5 percent cut in total U.S. carbon emissions that the Kyoto Climate Treaty would have required, and that the U.S. Senate unanimously rejected as far too burdensome in the 1990s.

Improved Energy Security

We know a long-distance runner who trains on a remote dirt road through the Angeles National Forest in the mountains of northern Los Angeles County. Although this is the most populous county in the United States, the runner says he rarely sees another person during the course of a two- or three-hour run. The only sign of civilization is an electric transmission line that crosses the ridges, carrying current from a central power plant in the Tehachapi Mountains to the city 60 miles to the south. One day in early summer 2008, the runner noticed that two signs he'd seen posted where the road passes one of the transmission towers were now lying in the dirt by the side of the road—apparently blown down by the high winds that frequent the mountains. He stopped to read one of them:

Southern California Edison, an Edison International Company
$1,000 Reward
This transmission line is patrolled.
Damaging these transmission facilities is a felony punishable by fine and imprisonment in the state prison.
1-800-455-6555

During the course of the summer, our runner passed under the power line every two or three days, and the signs remained on the ground. One night in late October, Santa Ana winds blasted the region, and the next day one of the signs was gone. The other had migrated a few feet and gotten snagged in chaparral. Months later, it was still there. During the entire summer and fall, the runner had never seen a Southern California Edison Company security vehicle or employee.

Although energy security is often considered synonymous with energy independence, there are important differences. Energy independence eliminates the threat of imported-oil or gas-supply lines being cut off by events such as the OPEC embargo, the oil fires of Operation Desert Storm, the Iraq War, or terrorist attacks on tankers in the Persian Gulf. But even the domestic supply of electric power is highly vulnerable in its existing form, due to its dependence on long-distance transmission. In addition to facilitating energy *independence* by reducing fuel demand, decentralization of electricity production will further heighten energy *security* in several respects:

- It will greatly reduce the vulnerability of domestic infrastructure to sabotage or theft. Aside from the question of whether terrorists ever decide to target central power plants or transmission lines, utilities are experiencing growing problems with thieves stealing wires and other components to sell as scrap metal. At last count, the United States had more than 500,000 miles of high-voltage transmission lines propped up by more than 2 million steel towers. It would be prohibitively expensive for all that infrastructure to be effectively patrolled, other than by automated systems that would add still more to T&D costs. Economically, it would be impossible to provide enough guards and vehicles, especially during a time of deepening economic travail. All the utilities can do is put up warning signs, such as the ones that lay rusting in the mountains north of Los Angeles.

- Decentralization also will reduce the risks of interruption by natural disaster. Power lines are vulnerable to wildfires, ice storms, tornadoes, and earthquakes. In summer 2008, a fallen power line started a massive fire in the San Fernando Valley of California, one of more than 2,000 wildfires that burned across the state that summer. With global warming, fires will likely become an ever greater problem, bringing down more power lines. Climate scientists say that even as warming brings worsening heat waves and drought, it will increase damages from extreme-weather events of all kinds. In January 2009, a massive ice storm knocked out enough power lines to leave a million homes and businesses in Kentucky and eight other states without power for more than five days, resulting in dozens of deaths. The costs of damages, the growing liabilities of utilities, and the monitoring costs cited previously can only add to the delivered costs of power from "central" facilities that aren't physically located near their customers. Decentralization will reduce the potential impacts of grid failure on local electric service.

• Decentralization will provide more redundancy per dollar of capital cost. The North American Electric Reliability Council (NERC) has determined that central power plants and transmission lines require an 18 percent redundancy capacity to be reliable at peak-load times, such as 105°F days when everyone's air-conditioning is on. As we noted earlier, that requirement adds substantially to the capital cost. A decentralized system needs only 3–5 percent redundancy. When Hurricane Katrina struck the Gulf Coast in 2005, the electric grid went down, notwithstanding its 18 percent extra capacity, jeopardizing the safety of people in hospitals and nursing homes whose lives depended on electricity-powered medical equipment. In Jackson, Mississippi, only one hospital remained operative for the first 52 hours—the Missouri Baptist Medical Center, which was powered by its own CHP unit. In the aftermath of the hurricane, much of the blame for lost lives was placed on the inadequacy of the levees and the slowness of the federal disaster response. Not mentioned was the set of archaic laws that prevented decentralized power, such as that at Missouri Baptist, from being used more widely—as it is in the Netherlands, Denmark, Finland, and even China.

With all these advantages, why hasn't movement been made toward the decentralization of electric power generation—not even a public discussion? Why isn't this a major political issue?

When things go wrong in governance, it seems that public debate often devolves toward ideological battle. Especially in energy and

climate policy, a running battle in the United States concerns government regulation. As we noted in our discussion of economic-growth
theory in previous chapters, free-market ideology holds that all government regulation impairs the productivity of the market economy.
Despite the catastrophic collapse of the deregulated financial markets in 2008, political conservatives remain wedded to the doctrine
that "markets always know best."[5]

Our view, however, is that government regulation is neither
inherently good nor bad—it can be either supportive to human life,
liberty, and the pursuit of happiness, or obstructive, depending on
how it's designed and managed. In the case of the Wall Street traders
and banks, the lesson of 2008—and a major factor in the election of
Barack Obama—was that the radical deregulation of financial markets and banks in the 1980s had been a mistake, and the Democrats
would now impose stricter controls. The media tended to depict this
as a reactive swing of the ideological pendulum and gave little attention to *how* the new re-regulation should be done.

In the case of the financial meltdown, key parts of good governance—transparency and accountability—were missing. In the
electric-power sector, it's just the opposite: Obstructive parts are
locked in place, like corroded gears. In every state, longstanding
laws prevent the production or sale of electricity from taking place in
a competitive market.[6] Reforming these laws to enable truly free
competition would provide an incentive to implement greater efficiency by means such as CHP and lower transmission and distribution costs. These changes would enable utilities, or their
competitors, to reduce the price of power to consumers. That would
achieve the "double dividend" of reducing fossil-fuel consumption and
leveraging the lowered price of energy to stimulate faster economic

---

[5] Consider this joke told by an economist, who cited it as originally published in
the *Wharton Journal*: "How many conservative economists does it take to
change a light bulb? Answer: None. They're all waiting for the unseen hand of
the market to correct the lighting disequilibrium."

[6] Utility people insist that competition exists, but they're referring only to the regulated competition allowed in some circumstances between different central
plants using the same obsolescent technology (in effect, a form of price-fixing),
not to new businesses that offer cheaper service to consumers and a far better
deal for the nation.

growth. This is one place in the machinery of the American economy where careful *deregulation* is critically needed.

Beyond the economic stimulus of a large infusion of cheaper energy, this strategy should yield some savings in capital costs, which will put the country in a far stronger position to (1) accelerate the ramping-up of renewable-energy industries, and (2) respond more effectively to the climate disasters we know are coming.[7] How? The electric-power sector isn't just a fixed infrastructure sitting at the foundation of the economy, like a stone wall that will still look the same 50 years from now. Power plants depreciate and must be replaced, and as population grows, new facilities need to be built—replaced either by "more-of-the-same" obsolescent technology (as hoped for by the Coal Rush bosses) or by something far more productive.

A recent study by the World Alliance for Decentralized Energy (WADE) estimated how much extra worldwide capital investment will be needed to keep building central generation plants instead of meeting new demand with local plants. Using the International Energy Agency base projection for 2030, WADE found that the world's energy industries have a choice of spending $10.8 trillion in capital costs of new central power plants in the next few years, or spending $5.8 trillion for decentralized generation to replace them. That estimated $5 trillion saving is roughly double what the Iraq War cost during its first five or six years. The U.S. share would be about a fifth of that, or $1 trillion. Decentralizing electric utilities could save the entire cost of the great American Recovery and Reinvestment Act of 2009—and probably do a better job of reviving long-term economic productivity.

Clearly, restructuring electric-power generation and distribution must be a principal girder for the energy bridge we envision. Although the full transition to decentralization and maximum feasible CHP might take 20–25 years, we'll need a strong policy commitment to launch it. That entails a greater willingness of government to

---

[7] In more prosperous times, California and other states had "rainy day" funds in their budgets. Every state now needs not just a rainy day fund, but a megadisaster fund. Such a fund is no longer a luxury, but a high-priority necessity even in the worst of times.

confront an entrenched industry that it hasn't previously confronted. The wider business community needs to look at the economic potential instead of continuing to reflexively avoid challenging an obsolescent system that has lost its economic punch. You might think this opportunity—a virtual domestic Saudi Arabia of free, clean, new energy—would be a no-brainer for the whole political spectrum. Deregulation is deeply appealing to conservatives (except perhaps to those who are profiting from the utility monopolies), and the boost to economic growth—and capacity to help reduce the federal budget deficit—should be attractive to both the political right and the left. Yet on those occasions when the central-power regime has been challenged, most countries have repeatedly rejected decentralization. Evidently, this is another area where we're in for a protracted fight.

Muscled by the Monopoly

In the late 1990s, the Massachusetts-based Cabot Corporation was the largest U.S. producer of carbon black, a petroleum-based product used mainly in making tires. But the company was being hobbled by a worrisome pattern, which was to become increasingly familiar to U.S. industry during the following decade: Although its foreign operations were thriving, its domestic production was only marginally profitable. The situation significantly worsened in 1999, as crude oil prices rose and rising imports of cheap foreign tires undercut the company's sales. Cabot needed a way to reduce costs. It was also under some pressure to reduce its air pollution, because making carbon black is an exceptionally dirty process that involves spraying oil particles into a flame. At its two Louisiana plants, Canal and Ville Platte, Cabot was emitting about 50 million pounds of pollutants per year from its smokestacks.

A possible solution came in the form of a proposal by a CHP engineering company, Primary Energy, to build a facility that would intercept Cabot's hot-flue gas and convert it to clean electric power, much like that being generated at the Mittal Steel coking plant described in Chapter 2. The agreement specified that the energy-recycling facility would be built next to the carbon-black plant, which was emitting enough hot gas to generate 30MW of electricity. The Cabot plant was using 10MW to produce its carbon black, and it was currently purchasing that 10MW from the local electric utility, CLECO, at a price of $55 per megawatt-hour (MWh). In the proposed deal with the energy-recycling company, the recycling facility would provide the needed 10MW for $45 per megawatt-hour. By using its own waste to power its plant (waste out one door, power back in the other), Cabot would gain multiple dividends: It would cut the cost of its Louisiana operation, improve the marginal profitability of its domestic production, and greatly reduce its embarrassing pollution output.

Because the recycling facility had a 30MW capacity and would be selling just 10MW back to Cabot, it would need another buyer for the remaining 20MW. Because the recycling facility wouldn't need to purchase fuel to make power the way the utility did, it could sell power at a discount (as it would be doing for Cabot). Every buyer likes a discount, so the deal seemed close to a slam-dunk. Another carbon-black plant, Columbian Chemicals (a subsidiary of Phelps-Dodge), was located just across the road. Like Cabot, Columbian was currently buying its power from the utility for $55 per megawatt-hour, and would be more than happy to buy it from the recycling facility for less.

However, there was one problem: The law in Louisiana—and almost everywhere else in the country—doesn't permit anyone other than the utility monopoly to sell power. It was made clear that the deal could work only if the utility served as a middleman—for the recycling facility to send its extra 20MW to the utility, which would then sell it to the buyer. Because the power would essentially go in one door of the utility and out the other, with no intermediate processing, the utility would incur only an administrative cost, which would be negligible. Ideally, the recycling facility could sell to the utility at $45 and the utility could sell to the buyer at $50—still passing along part of the discount the buyer would have had if the law hadn't prohibited a direct transmission across the road. You might think that CLECO would accept a windfall profit of about $5 per megawatt-hour for itself while helping perform the public service of substantially reducing carbon emissions and other pollutants.

However, CLECO saw it differently. They viewed it as a loss of retail sales to both Cabot and Columbia Chemicals. The deal also undermined CLECO's pending case to the public-utility commission for building a new $50 million transmission line to serve the Canal and Ville Platte area, the cost of which would then be added to their rate base for the whole state. But the real problem with approving a recycling deal was that, by reducing demand for power in that area of the state, the recycling facility would also undermine the utility's case for building future "central" power plants, the cost of which would also be included in its rate base. Why go along with a project that would reduce the state's fuel use, when it could hold out for an arrangement that would let it profit from *increasing* fuel use?

Instead of agreeing to the opportunity for a win-win solution for the community, the utility said it would pay only $20 per megawatt-hour for the remaining 20MW—a deal killer, because the recycling facility couldn't operate profitably at that rate and clearly couldn't undertake construction if that was all it would get. CLECO stuck to that price for a year before raising its offer to $28—still far too low for the plan to work. At this point, the Louisiana governor was brought in and the civic benefits (including the new employment the proposed facility would bring) were stressed. After two more years the utility raised its offer to $38, which was close but still tauntingly short of what would have worked. By that time, however, both Cabot and Primary Energy had reached the end of their rope, and the project was abandoned.

Soon afterward, CLECO applied for permission to build its hoped-for new central power plant, partly to provide the capacity that the energy-recycling plant would have provided. However, an additional supply of fossil fuel burned at 33 percent efficiency, along with its commensurate power-plant emissions, would now be required. In the years since then, the Cabot facility has continued pouring its emissions into the sky instead of having them intercepted and turned into clean power. Bottom line: CLECO was able to prevent the construction of a 50MW facility that could have produced more than 400,000MWh per year of clean energy while substantially reducing pollution.

The carbon-black case is an example not only of a lost opportunity for movement toward energy independence and reduced carbon emissions, but of apparently deliberate obstruction of that movement. Yet it's a common operating procedure nationwide: blocking a strategy that would be vastly easier and more productive than using military muscle to increase the supply of oil from Iraq or Venezuela, for example. The scandal is that the power plant waste stream remains turned on, night and day, like a forgotten garden hose left on in a drought. Federal and state laws, along with entrenched practices and preconceptions, perpetuate this endless waste—and its contribution to climate change—by protecting the power companies' monopolistic control and wealth. Investors, executives, lobbyists, and compliant utility commissions keep the consequences out of the media and public consciousness, thereby persuading Americans that the real need is to increase fossil-fuel supply. To extend the forgotten-hose metaphor, it's like spending tax dollars to increase the subsidized water supply to homes instead of turning off the hoses when the garden isn't being watered.

In our efforts to penetrate this thicket of paralyzing laws and practices, we found it would take an encyclopedic volume to track them all. But here, in brief, are the main obstructions:

- **Laws against selling electric power to third parties, and laws that prevent privately owned wires from crossing public streets**—An energy-recycling plant can go into business turning waste heat into power only if it can sell back to the monopoly utility at a profit, or (in some states) if it has a nearby customer on the same side of the street. If it's a company such as Mittal Steel, which can use the waste from its coke plant to provide power for its adjacent steel plant, it's legal. But otherwise, it's no go. All 50 states have laws making it illegal to cross the streets with wires carrying electricity from anyone other than the area utility; in a number of states (such as Louisiana), it's even illegal to sell electricity to *anyone* but the utility. Never mind U.S. anti-monopoly laws—the electric-power industry has a monopoly in every city or town. And on every street.

- **Two sets of books determining electricity rates**—One is for the central power plants and the other is for any aspiring competitors from outside the system. Ostensibly, the rate an energy-recycling company can charge the utility is determined by the "base" rate the state utility commission allows the central power plant. However, the power company is also allowed to add to its base rate the very large costs of transmission and distribution (T&D)—all those hundreds of miles of power lines. Moreover, its infrastructure has to be built to accommodate peak loads, which are much higher than the average load, and which allow the power company to raise the price even more. However, the energy-recycling companies or other local competitors don't need long-distance infrastructure and would be able to sell electricity to local consumers at a substantial discount—but they're not allowed to. The laws prohibit the recycling plants from taking advantage of the fact that they've found a more economical way (local generation) to provide electricity. For them, the "free market" is illegal. And as a result of just that boondoggle alone, a vast amount of America's purchased energy is thrown away.

- **Government subsidies for the least-efficient systems**—As we noted earlier, a combined heat and power (CHP) plant can reach 50–80 percent efficiency, compared with only 33 percent for a central plant. Yet the central plants that are part of state- or municipality-owned systems pay no income taxes. As a result, consumers get hoodwinked by another deft shuffle: Politicians keep consumers relatively pacified by bringing them seemingly cheap electricity, while making them pay a hidden premium for it by increasing their taxes or fees to subsidize the utility's share of the burden.

- **Exemptions from predatory practice laws**—U.S. antitrust laws prohibit the practice of "product bundling," in which a large company offers a big incentive to buy one of its products by packaging it with another product that it can sell at a loss—at a price that competitors for that second product have no chance of matching, and that can help drive those competitors out of business. For example, a few years ago, Kodak offered ultralow prices on copiers to consumers who also purchased maintenance agreements. The U.S. Supreme Court ruled that this violated the antitrust rules against bundling. Yet when electric utilities offer discounts for electric lights in buildings if the consumer also agrees to buy electric heating and cooling, there is no challenge. All-electric discounts thus discourage the use of non–fossil fuel systems, such as solar panels.

The multiple benefits of decentralization make this monumental shift in energy production not only essential, but inevitable. Old coal-burning plants will depreciate, and as the American economy becomes increasingly electrified (with the coming of electric cars, for example), U.S. demand for new electric-generation capacity will continue to grow even as energy intensity continues to decline. The opportunity to cut U.S. fuel use and emissions by 15–20 percent, cut capital costs by 50 percent, and vastly increase energy security—all while maintaining energy services at present levels—will be given further impetus by the continuing "technological progress" of the micropower revolution. As with computers, electric-power generation will get smaller, quicker, and cleaner—and cheaper. Sooner or later, the big central power plants, like the Soviet-era apartment blocks and the "central planning" that built them, will come to be recognized as obsolete, inefficient, and ugly. To serve as a key girder to the transitional bridge, the first stage of that recognition needs to happen sooner, not later.

6

Liquid Fuels: The Hard Reality

The American love–hate relationship with cars has made motor vehicles the main focus of energy issues in the public consciousness. That has created a precariously misleading picture of the overall energy challenge, because it distracts policymakers from the main needs of the transitional bridge.

That's not to say motor vehicles aren't central to American life, culture, economics, and environmental decline. They are. The anguished debate about whether the U.S. "Big Three" automakers should be bailed out made that clear. But that debate was also a huge diversion from a more fundamental reality: There's no future in spending a large part of our limited time and resources making further refinements in the obsolescent technologies of the internal combustion engine, the two-ton personal vehicle it powers, and the gasoline that fuels it. As noted in previous chapters, a typical car in the United States with one person in it gets about 1 percent payload efficiency. Making the gigantic investment needed to shift all cars to hybrid or electric would bump that up to no more than 2–3 percent. If we're resigned to a lifestyle that requires moving a ton or two of steel every time we move a person, we can claim that the car's energy efficiency for that task is 10–20 percent, but that still wastes eight or nine of every ten barrels of oil used for making gasoline in a world where oil is about to get increasingly scarce.

Meanwhile, the impact that the early romance with cars had on the shaping of the American economy (the building of roads and sprawling suburbs, the political dominance of the auto industry, and the century-long suppression of public transit) led to the same pattern in cargo transportation: Trucks and aircraft reduced the use of trains, and the once-thriving railroad system fell into disrepair.

It's too late to walk away from cars now (except literally, in some important but limited ways we discuss later). We'll need to design the cities of the late-twenty-first or twenty-second centuries to enable most people to live without individual two-ton car ownership and with much less car dependence. But there's no way we can do more than make the first steps toward that future during the transitional period of the energy bridge. America's dilemma is that it must achieve huge fuel reductions in the next two decades while continuing to get by with a motor vehicle–dominated transport system that is inherently inefficient, even at its best.

Americans have cooled somewhat in their ardor for cars—thanks to the stresses of worsening congestion, family-wrecking accidents[1], and lung-destroying air pollution—but the toxic romance is now flourishing in other countries. For example, China has a population four times that of the United States, and the growth of car ownership is just getting rolling. The United States has reached car saturation, with more than one car for every two people, but in China, the ratio is one car for every 300 people. With their huge trade surplus in dollars, if the Chinese get anywhere near the U.S. ratio of cars to population, with anything near the same rates of fuel use and carbon emissions, they won't have enough land for the highways or fuel for the engines. Even if they did have enough land or fuel, the impact on climate would be unthinkable

Although mainstream expertise has now come around to Al Gore's once-ridiculed assertion that the internal combustion engine (ICE) needs to be phased out, the most immediate issue is not the engine—it's the fuel. Further refining ICE technology, which is very near the end of its road, won't significantly improve fuel efficiency. We can achieve higher automotive fuel efficiency by making cars lighter and more reliant on electricity instead of petroleum. But to keep it real, remember that electric cars aren't free of fossil fuel. Making the electricity currently uses great quantities of coal and

---

[1] In 2006, a typical year, 5,973,000 motor vehicle crashes occurred in the United States—one crash for every 52 people. In those crashes, 2,575,000 people were injured and 41,059 were killed. For every American soldier who died each year in the first five years of the Iraq War, 50 Americans died in car crashes back home.

natural gas. And because we're talking about an energy bridge that must begin working in a relatively short time, remember that most of the Ford pickups, Dodge Rams, Cadillac Escalades, Toyota Tacomas, and Hummer H2s that are being purchased today will still be on the road 8–10 years from now—still burning a lot of gasoline.

Transportation in the United States is utterly dependent on the availability of liquid fuels. Politicians who pretend that more nuclear power plants will somehow solve the energy crisis, which is an oil crisis, are misleading the public and probably themselves. All transportation modes, with the sole exception of electrified railways and trams, depend upon mobile power sources. And all the mobile power sources we know of, with the still minor exceptions of electric batteries and fuel cells, are internal combustion engines using liquids derived from petroleum—gasoline, kerosene, or diesel oil—as fuel. Electricity, whether from nuclear plants, solar farms, or wind turbines, won't create liquid fuels.

A hundred years ago, railways transported virtually all goods and most intercity passengers and commuters, and most railways used coal as a fuel for their steam-powered locomotives. Those locomotives were noisy, smoky, and thermodynamically inefficient, because coal was so cheap that nobody had much incentive to redesign the engines. Starting in the 1930s, the diesel-electric combination tripled the efficiency of the old steamers, but that innovation kept the U.S. railways competitive only for goods shipments in a few long-haul, intercity markets. In virtually all other transport markets, motor vehicles won the day, especially after World War II, when burgeoning demand for motor vehicles created a demand for new highways—which the government built and paid for. By contrast, the privately owned railways still had to operate and maintain their own tracks and rights-of-way. It's no wonder that socialist governments after World War II nationalized most railroad systems outside the United States.

In the United States, where socialism was anathema, the previously efficient and formidable railway system was simply allowed to wither away. General Motors led a campaign to remove urban tramlines, leaving the market for intra-urban travel to diesel buses (supplied by GM) and private cars (also made by GM, Ford, and Chrysler.) The creation of Amtrak was a late attempt to stem the decline of railways, but the magnitude of investment that would be needed to create a modern network of high-speed passenger trains, such as Europe and Japan are now creating, has not been forthcoming. (It's impossible to run modern high-speed trains over tracks designed for nineteenth-century trains. Sharp curves must be eliminated, hills must be flattened, and the entire roadbed must be relaid to higher standards.) Attempts to "leapfrog" the high-speed railways being built elsewhere by introducing ultra-high-speed maglev trains in Florida, Nevada, and elsewhere have failed, largely because of no government support. The prevailing theory has been that if the technology is really justified, the private sector will finance it. (In the United States, the airlines—utilizing military-developed jet engines and government-funded airports and air-traffic controls—have largely dominated the growing market for intercity and international travel.)

Substituting automobile and truck transport for rail transport has reorganized the urban landscape in ways that will be almost impossible to reverse. U.S. cities are spread out over enormous areas that are extremely difficult to serve by public transportation. It's estimated that half of the entire area of Los Angeles is devoted to highways, streets, driveways, and parking lots. Most of that paved area is dark in color, and its absorption of heat contributes significantly to local climate warming (the "heat-island" effect), which then increases the demand for air-conditioning that increases energy consumption even more. The enormously increased petroleum consumption associated with urban transportation in decentralized cities such as Los Angeles has unquestionably aggravated the climate problem. Most of the blame deservedly goes to the single suburban commuter driving a 2-ton SUV, because millions of solo drivers are consuming fuel. Short-haul air transportation, another major polluter and liquid-fuel user, is growing even faster. What can we do to change this? The solution usually advocated by politicians is "new technology." We need to take a good look at the new technologies now being proposed.

Corn Ethanol: A Fuel's Errand

A good starting point is the one that has advanced most rapidly during the past few years in parts of the United States—the partial substitution of ethanol (ethyl alcohol) for gasoline. Gas stations sell an unobtrusive mixture of the two fuels, in which the ethanol fraction is rather small. This mixture makes some sense from the retailer's standpoint because ethanol contains less energy (exergy) per gallon than gasoline. A vehicle using pure ethanol either would have a much shorter range per refill or would need a fuel tank that was about twice as big. But if the fuel is only 10–15 percent ethanol, the difference in range is much less noticeable.

Ethanol is made essentially the same way as an alcoholic beverage—by fermentation of sugars from agricultural crops. Beer is made from barley and hops, wine from grapes, and ethanol from corn. In 2005, U.S. ethanol production reached 4.5 billion gallons. It took more than 14 percent of the nation's corn harvest, but it replaced only 1.7 percent of the nation's gasoline energy. In the same year, biodiesel from soybeans consumed 1.5 percent of the soybean harvest and replaced 0.09 percent of the diesel fuel that trucks and off-road machinery consume.

However, that's not the whole story. Producing biofuels also consumes fossil fuels for processing and transportation. So if the entire U.S. corn crop were converted to ethanol, the aggregate reduction in total fossil fuel (gasoline) energy consumption would be just 2.4 percent. If the entire soybean crop were converted, the saving would be 2.9 percent. Considering that we still want to have our corn flakes, pork chops (corn is used to feed hogs), cooking oil (both corn and soy), or soy-based baby formula, any realistic level of corn or soy consumption for fuels would do virtually nothing to help end U.S. dependence on foreign oil.

Environmentalists have depicted corn ethanol as an outrage, and it's worth a hard look at what they are saying—particularly because Barack Obama, whom environmentalists overwhelmingly supported during his presidential campaign, was an unabashed advocate of increased ethanol production. Either he knew something they didn't or (as we suspect), for all his remarkable education and willingness to hear opposing points of view, he didn't know the hard facts about ethanol. Or given the political debt he owed to Iowa (where his campaign was so dramatically launched), perhaps he couldn't afford—at that time—to raise troubling questions about corn subsidies.

That's not to say that after the election was won, the Obama administration could turn its back on the Midwestern farmers whose support for him was based not just on ethanol, but on the larger promise of revitalizing the U.S. economy. Energy issues are more complex than either of the 2008 major presidential candidates could afford to acknowledge in stump speeches, or in a media environment of ten-second sound bites. But as we have argued in this book, a truly effective economic revival can be achieved only by reducing the cost of energy services. As a chorus of environmental groups pointed out in protest of the Bush policies, corn ethanol doesn't do that.

In fact, as the corn subsidies began flowing into the big agribusiness corporations that were the main beneficiaries, the media bristled with objections—in *Rolling Stone:* "The Ethanol Scam: One of America's Biggest Boondoggles" (August 9, 2007); in *Slate:* "The Great Corn Con" (June 26, 2007); and a cover story in *Time:* "The Clean Energy Scam" (May 27, 2008). One widely cited objection was that if you count not just the processing and transporting of ethanol (as the U.S. Academy of Sciences calculated), but also the energy used to grow the corn (in the form of agricultural chemicals, fuel for tractors, and so on), corn ethanol actually takes more energy to produce than it provides to the national supply at the gas pump. Is that true? The Natural Resources Defense Council investigated that question in a review of six studies of "energy return on investment" (EROI). The results of the investigation suggested that the view conveyed by *Time* and others might be exaggerated, but only a little. Five of the six studies found positive EROI ranging from 1.29 to 1.65 (129 to 165 percent), depending on assumptions—meaning that the energy contained in the ethanol is *slightly* greater than the energy required for the farming, harvesting, processing, and distribution of it. The sixth study reviewed by the NRDC was much more pessimistic; it concluded that ethanol from corn contains less energy than the inputs needed to produce and distribute it. By comparison, the U.S. EROI for domestic petroleum is approximately 15. In the 1930s, the energy return was more than *100*. We can conclude that ethanol adds either very little or nothing to the U.S. energy supply, considering the energy needed to produce it.

Despite that reality, the Bush administration advocated increasing U.S. ethanol production to a massive 35 billion gallons, of which 15 billion gallons were to come from corn and the rest from other sources. If carried out, this plan would dramatically reduce production of food, especially meat, because hogs and steers raised for meat are largely corn-fed. Moreover, the biofuels are not *cheaper* to produce than gasoline or diesel fuel, so in addition to their lack of contribution to the energy supply, they will do nothing for economic growth. The wholesale cost of a gasoline gallon equivalent (GGE) of ethanol in 2005 was $1.74, compared to $1.67 for gasoline. It was $2.08 for a GGE of biodiesel, compared to $1.74 for diesel oil.

Despite being an economic sinkhole, ethanol is being promoted with the help of generous federal subsidies of 76¢ per GGE for corn ethanol and $1.10 per GGE for biodiesel. In addition, SUVs get mileage credit for the fraction of gasoline or diesel fuel replaced by ethanol. The automakers are required to report their fuel economy only with respect to the gasoline consumed—another win for the SUV lobbyists.

However, the case for corn-based ethanol is even weaker than those facts suggest. This analysis doesn't include irrigation requirements (see Chapter 9, "The Energy-Water Connection"), nitrogen pollution (from excess fertilizer use), or the expanding eutrophication (dead zones from farm runoff) in the Gulf of Mexico, nor does it consider that agricultural subsidies also create incentives to clear forest to create more cropland. When alternative uses of land are considered using reasonable assumptions, the *opposite* strategy—converting cropland to forest—would more effectively reduce global greenhouse gas concentrations in the atmosphere (by sequestering carbon in biomass) than converting forest to cropland used to grow corn or soybeans.

Looking at the big picture of what's required to build a viable energy bridge to the sustainable economic future, it's obvious that pouring large federal subsidies into a solution that produces very marginal, if any, energy-supply gains isn't a viable solution. We stressed in previous chapters that for the transition bridge to work, it must quickly increase supply (or efficiency, which is de facto supply) at significantly reduced, or even *negative*, cost.

A Fuel for the Future—Beyond the Bridge

Clearly, corn-based ethanol and biodiesel from soy beans don't qualify as economical or negative-cost solutions. A rude characterization of the ethanol program is that it's a costly government handout for large campaign contributors and farmers who participate in the Iowa primary electoral caucuses. The most that can be said for these fuels is that they might become an inadvertent first step toward a somewhat better solution—*cellulosic* ethanol, from woody crops or cellulose wastes (such as corn stalks). Researchers are also exploring the possibility of making ethanol from grasses grown on marginal lands that aren't usable for food crops. But the technologies for these forms of ethanol aren't fully developed and no commercial production is currently occurring. Even if commercial production begins in the years ahead, it will take years more (as in the critical cases of solar and wind) to get up to scale.

An abundance of visionary writing has explored the possibilities of the post–fossil fuel future. This book focuses on a more prosaic subject—the near-term challenges of getting from here to there. In the particularly challenging area of liquid fuels, while none of the ethanol options will help build the transitional bridge, it's important to distinguish between those that might warrant continued R&D for possible uses in the post-bridge future and those that are wasting valuable resources at a critical time. Keep in mind that although ethanol is touted as a substitute for fossil fuel, it still emits carbon dioxide when burned, just as gasoline does. On the other hand, if future needs include moderate quantities of industrial alcohols, continued research on the cellulosic ethanol processes might be justified if they don't occupy food-producing land. For powering motor vehicles or aircraft, these biofuels have no future.

For the foreseeable future, the best liquid fuel option might be diesel oil, from either plant crops or other sources. European cars get much better fuel economy than American cars. One reason is that Europe has more small cars than the United States does, but a less familiar explanation is that about half of the cars sold in Europe have diesel engines. Diesel engines get about 50 percent fuel-burn efficiency, compared with 27 percent for gasoline engines. Diesel cars also have a reputation for being more polluting and have therefore made little headway in the United States. But technical improvements in recent years suggest that providing incentives to shift more car purchases to diesel (for example, by setting tax rates lower for diesel fuel than for gasoline), could make significant gains in U.S. fuel economy.

Fuel for Airplanes

Among the technological "white elephants" we are stuck with as we prepare to enter a very difficult energy age, the GM Hummer has acquired a reputation that it might not live down in a hundred years. Like the Ford Edsel a half-century ago, the Hummer has gained a place in social history that its manufacturers never intended.[2] But in the big picture of our present liquid-fuels problem, the real elephant in the living room isn't the one that's been sitting in the driveway; it's the one on the *runway*—the jet plane.

Airplanes burn about 2 million barrels of oil per day in the United States—roughly 10 percent of total U.S. liquid-fuel consumption. But that modest percentage is misleading, because it masks how much is used per passenger-mile (or suitcase-mile, or Chilean sea bass–mile). A typical commercial airliner consumes two to three times as much fuel per passenger mile as a car, or about the same as the biggest Hummer model. When you and 250 other passengers step aboard an airliner to fly from San Diego to Washington, D.C., the fuel you and your fellow travelers are about to consume is the same as if everyone tore up their plane tickets and instead drove 250 Hummers across the country.

The problems with aircraft are more intractable than those with terrestrial vehicles. Aircraft have higher payload efficiency than cars: Passengers account for about a third of the weight of a loaded plane. But fuel accounts for another third, so planes are inherently much more energy intensive. Most important, planes don't have alternative (nonliquid)-energy sources to begin phasing in, the way cars and trucks do. Cars have at least begun the long transition to gas–electric hybrids, plug-in electrics, and hydrogen fuel cells. Those technologies can't fly planes, at least not anytime soon.

---

[2] Environmentalist criticisms of oversized SUVs—of which the Hummer was the most notorious—began routinely referring to them in recent years not only as "gas hogs" but as wannabe military vehicles that their owners could use to intimidate other drivers. In 2009, a Honda TV commercial satirized that "muscular but efficiency-challenged" reputation of the biggest SUVs by depicting an out-of-gas SUV driver being given a ride by a Honda driver. The Honda driver makes it clear that *he* doesn't have to worry about running out of gas, because he's driving such a fuel-efficient vehicle. The frustrated SUV driver, sitting in the back seat, petulantly replies, "But can it crush cars?"

Lots of chatter on the Internet has circulated about alternative fuels for planes. The president of Virgin Atlantic, Richard Branson, attracted a lot of publicity in 2007 by announcing the first successful flight of a commercial airliner using biofuels. But experimental prototypes, whether of aircraft or fuels, are far too costly to be competitive in their early stages of development. Even if they are surprisingly successful, it will take years to bring them to market. And if the jet biofuels are similar to corn ethanol (which doesn't work for airliners because it freezes at their cruising altitudes), they'll require about as much energy to produce as they bring to the fuel tank.

Possibilities on the horizon at least offer hope of keeping some planes flying after oil is no longer affordable. Hydrogen or liquefied natural gas might someday be viable. There is also the prospect of making a biofuel from the algae that grows on ponds. California's Solazyme Corp. has produced a form of kerosene from algae grown in tanks, which reportedly meets most of the technical standards required for commercial jet fuel—including not freezing at 30,000 feet. The pond-scum solution might seem as unlikely now as the idea of a fish farm once did, but fish farms are now a major industry, with a lot of the former cotton farms of the Old South now raising fillets instead. And a major advantage of pond scum compared to corn ethanol is that it doesn't steal food from the mouths of our children's children. But like cellulosic ethanol or biodiesel, pond scum still produces carbon dioxide emissions when burned. It might alleviate some of the economic pain of post–peak oil, but it won't help mitigate global warming. And it won't be ready in time to help with the energy bridge.

The hard reality is that no new liquid fuels in the developmental pipeline can make a material difference to the transitional economy. The one strategy that *can* make a difference is restructuring the existing energy economy so that we use the supply of petroleum we already have far more efficiently. Like it or not, our liquid fuel for the next decade or longer will be a petroleum derivative. We can stretch the supply to meet the minimal need for motor vehicles and aircraft through a combination of cultural and policy changes. These include more demanding corporate average fuel efficiency (CAFE) standards for planes as well as autos, and a range of changes that reduce the number of miles traveled. We discuss these topics in the next chapter,

"Vehicles: The End of the Affair," and in Chapter 8, "Preparing Cities for the Perfect Storm."

Over the next decade, some help can come from companies redesigning airplanes to be substantially more fuel efficient. Boeing's new Dreamliner is expected to use 20 percent less jet fuel than its older models, thanks to a more aerodynamic shape and lighter weight. The new planes will replace large parts of the aluminum body with carbon-titanium composites, which are both lighter and stronger than aluminum, and which can eliminate the need for about a million rivets, screws, and other parts per plane. But redesigning, testing, and ramping up production of new airliners or military airplanes in this way, like building oil platforms or nuclear power plants, can take a decade. In the meantime, both we and our military might need to cut back sharply on the sheer volume of hauling heavy objects through the air that we think is necessary. As the supply of petroleum declines and the price rises higher than ever, the main task won't be to keep trying to replace gasoline or jet fuel with new fuels that compete with food crops for our land, but to learn to use the remaining fossil fuels far more efficiently.

7

Vehicles: The End of the Affair

As the economic crisis rapidly escalated in 2008, commentators raised a question that would have been unthinkable a few years earlier: whether the "Big Three" American automobile manufacturers deserved to survive. The chief executive officers of General Motors, Ford, and Chrysler had all come to Washington to plead for a government bailout, and Congress was in a truculent mood. It was hard to believe that the auto companies wouldn't get some kind of help, because so many businesses and jobs depended on them. But first the politicians and public were going to vent. Why had the American manufacturers *taken so long to wake up* to the realization that this was a different world than that of the 1950s or 1960s, when GM was the largest company in the world and Americans were agog with anticipation as the new models of cars rolled out each year? Why had GM finally introduced its EV-1 electric car in 1996, then dropped it three years later and cynically brought out the in-your-face, gas-swilling Hummer instead? Why had the U.S. manufacturers let the Japanese and Korean manufacturers not only get their feet in the door of the greatest American industry, but threaten to take it over? Why had the CEOs bearing their tin cups come to D.C. in corporate jets?

Never mind whether the CEOs "got" what the venting was about. Just the fact that such questions were raised might have signaled that the United States was indeed waking up to a clearer view: Cars were no longer the cultural icons—the ultimate symbols of freedom, aspiration, adventure, and success—they had once been. The luster hadn't entirely gone away, but younger people were enamored of other things now: flat-screen TVs, iPhones, MySpace, and video games. The people who had lived in a dreamland of drive-in movies and Route 29 cruising in their '61 Chevys or '56 Fairlanes with 202 horsepower Thunderbird V-8 engines were now retired or soon would be. For many, a car was now just a way to get from home to the mall or to the office. And that seemed to be getting more difficult by the day.

That's not to say motor vehicles were any less dominating of the landscape in the early twenty-first century than they had been a few decades earlier; they were now *more* dominating, and that was a big part of the problem. Congestion was now the rule, not the exception. For many, driving had gradually shifted from dream to recurrent nightmare. And since the environmental awakening of the 1970s, Americans had become generally more aware of the impacts of auto exhaust—first on smog-related health problems and the habitability of car-clogged cities, and then, in the 1990s, on global warming. Meanwhile, the new industries of telecommunications and digital entertainment had made their entrance. For the young, the focus of excitement had shifted from on-the-road or in-the-sky adventure to cyberadventure. The kids of the 1960s had wanted to be hot-car drivers or astronauts; those of the 1980s had wanted to watch Star Trek on TV; and the generation after that had wanted to rocket at hyper-speeds through cities, space, and time via Xbox or Wii. Hot rods couldn't compare.

The importance of that shift is more significant than it might appear, given the current car-clogged condition of U.S. cities and suburbs. A basic premise of this book is that the fossil-fuel economy has enormous momentum and will take time to change. But the dreams of the young—and their energy—will drive that change. And as we have noted throughout the book, when new markets arise, it's not the technologies they incorporate that make them hot; it's the services those technologies provide. Cars and trucks have historically provided core economic services, but at least some of those services can now be better provided by other means.

One of the keys to a more energy-intelligent economy is to make more emphatic distinctions between the different services cars and light trucks have historically provided. For a century, cheap gas and lack of environmental awareness have gotten us used to the idea of an all-purpose vehicle: the station-wagon of the 1950s, or, later, a van, SUV, or pickup truck that you could use for *whatever*—a thousand-mile vacation trip or a half-mile drive to the store for a six-pack.

Historians of the future might look back at that all-purpose ride and regard it as something comparable to a warlord using his sword for all cutting purposes—whether carving his meat, picking his teeth, or cutting off an enemy's head.

In the twenty-first century, separating the services of "highway" travel, intermediate-distance suburban or urban mobility, and short-distance mobility is essential to intelligent energy and climate management. Equally essential will be revisiting the idea of mobility itself. Are these different categories of mobility really the services we want? In some cases, they are, because we really do need or want to be physically transported from here to there. In other cases, maybe we don't. You need to buy things, but that doesn't necessarily mean you have to physically transport yourself (and your personal 2 tons of steel) somewhere to do so. You might indeed want to get to places such as the beach or a family reunion, where actually being there is what it's all about. But do you ever have a physical need to get into a Walmart store? Or do you just want some of the stuff that's *in* the big box, in which case, maybe it can come to you?

A Major Mental Hurdle

For twenty-first-century Americans, it's hard to separate the idea of progress from that of new technology. For a half-century, the neo-classical theory of economic growth has blurred those two concepts in its terminally vague notion of "technological progress" as the great exogenous driver.

Accordingly, Americans might reflexively expect that if we're to find a better means than conventional cars and SUVs of providing short-distance mobility, it will be some spectacular new technological breakthrough. (Moving sidewalks? Flying scooters?) We have a huge mental block to the idea that progress might come from the past, or from unappreciated regions of the present. Yet we've recently experienced numerous instances of such progress. Examples include the burgeoning market for organic food (what virtually all food was before the advent of pesticides); the new respect and growing markets for ancient medical techniques such as acupuncture; and architects' and designers' new respect for time-honored building materials—stone, wood, and terra cotta—instead of the once "new" materials, such as vinyl siding or linoleum floors, that seemed so modern when they first appeared.

That brings us to the much-ignored bicycle, which is the most widely used form of short-distance personal transport (other than walking) in the world today, and which in many respects is the best. Bicycle advocates are passionate, but they've been consistently brushed off or marginalized. In some regions, such as China, they have lost ground. But in an era of increasingly dysfunctional urban transport systems all over the world, their day might yet come.

China has more than 420 million ordinary human-powered bicycles, compared to about 32 million private cars. But the private car owners and high-level functionaries who are entitled to official cars in China are far more affluent and influential than the cyclists, and they are supported by a belief in high places that it's necessary for China to develop a big automobile industry to be fully industrialized—as dramatically demonstrated by a Chinese company's acquisition of Hummer from bankrupt GM in 2009. Thanks to their disproportionate affluence and influence, the car owners and drivers are compelling the government of China to build freeways where rice paddies used to be. The new freeways cut travel times between major airports, industrial centers, and hotels for the privileged few, but they simultaneously spill heavy automotive traffic onto all the connecting streets and roads. The inevitable result has been some of the planet's worst congestion and smog (recall athletes' worries about breathing the air in Beijing at the 2008 Olympics), making it much harder and riskier for local bicyclists to compete for road space.

However, in some European countries with more mature economies—most notably the Netherlands, but also in significant parts of Belgium, France, Germany, Italy, Spain, and Portugal—bicycles have become a major mode of transportation in towns and cities. Dutch commuters use bicycles to go to school, to shops and offices, to sports facilities, and to the nearest railroad station (which is never very far). In all these places, they find convenient bicycle parking spaces. The climate is by no means benign for much of the year, but it doesn't stop the cyclists.

Bicycles aren't really suitable for commuters wearing business clothes, parents with small children in tow, shoppers with big loads of groceries, or the elderly and infirm. However, the question isn't whether bicycles can replace cars in cities, but whether they can replace *some* automotive traffic—enough to visibly reduce a city's fuel consumption and carbon emissions. *Some* of the people can use bicycles for *some* of their trips, in ways that don't constitute hardship— and might be conducive to better health and enjoyment of the urban environment. The parent who needs a car (or other capacious mode) to carry the kids to soccer can still leave the car parked and use a bike for short, solo trips to the park or gym, or, in good weather even to work. If urban dwellers can shift some of their mobility needs to human-powered transport, their actions *can* significantly reduce the fossil-fuel use of the region. Segregated bicycle lanes, often in parks or along rivers or canals, are features of most European cities, but of only a few in North America. Starting in Amsterdam in the 1960s, several large European cities have experimented with free "bicycle sharing." Most of these plans were inadequately thought through or too small in scale to succeed, but the plans are getting more sophisticated.

Paris offers a promising example. The city has set aside special lanes on major boulevards for buses, taxis, and bicycles. (The premise is that professional drivers can share road space with cyclists without endangering them.) In July 2007, the city inaugurated a program called Velib, with an initial endowment of 10,600 bicycles of uniform design (paid for by Cyclocity, which is a subsidiary of the big advertising firm JCDecaux) and allocated them to 750 reserved parking racks around the central city. Within the next year, they increased the number of bicycles to 20,600, and the number of parking stations to 1,450. Credit cards activate the bikes, which are electronically monitored. Users must pay a small annual fee to belong to the Velib Club. The first half-hour of each trip is free, with a nominal hourly rental fee thereafter.

In Lyons, France, where a similar system was introduced in 2005, each bike is used an average of 12 times each day, and 95 percent of the trips are free. Most trips are point-to-point, between one reserved parking place and another. We don't yet have good data on the impact of these shared bicycles on other modes of travel; it's questionable whether they're replacing many auto trips, because very few Parisians or residents of other bike-friendly European cities use cars for short trips inside the city. The shared bicycles mainly replace walking trips or bus and metro trips, reducing congestion on those modes. The main benefit to users is speed—the bicycles are faster than taking the bus or the metro between most pairs of destinations. We found little evidence of energy conservation or cost saving.

However, the shared bicycle programs in Paris, Lyons, Vienna, and elsewhere might be stepping stones to a more significant near-future program—shared electric vehicles aimed at reducing commuter trips from within the city or the inner suburbs, and eventually from the outer suburbs, where most commuters travel by private car.

Next Step: E-Bikes

The average time or distance for a conventional bicycle user is currently less than half an hour or 5 kilometers (3 miles), enough for some commuters to get to work and for many others to get to a public-transit station. Many cyclists can (and do) go much faster and farther, at least where the terrain is fairly level. However, we currently consider those individuals trailblazers. Where caloric energy is now required, solar or wind energy will provide an easier ride in the future.

The next step beyond the human-powered bicycle (not displacing it, but greatly augmenting it) will be the battery-equipped electric bike, or "e-bike," which is capable of an average speed of 9–13 miles per hour, depending on traffic. China already has at least 30 million e-bikes, out of the total bicycle population of 450 million. The market for e-bikes in China was just 40,000 units in 1998, but by 2006, it had exploded to an estimated 16–18 million units. More than 2,000 mostly small companies produce e-bikes.

Electric bicycles and scooters are still rare in Europe and America, but they have the potential to change the commuter game radically, even in a spread-out American metro area such as Los Angeles. Although gasoline-powered motorcycles have long been a part of the vehicle mix and get much better gas mileage than cars, large barriers stand in the way of their wider use. One barrier is the risk of accidents, which is much higher for motorcycles than for cars—even though the risk of driving cars is higher than almost anything else that most people do. Many people seem to have reacted to motor vehicle safety concerns in the opposite direction, choosing to drive heavy, fuel-hungry SUVs precisely because (encouraged by auto company marketing) they believe big vehicles are safer.

Gasoline-powered motorcycles can't be a part of the transition to more environmentally benign urban transportation because of their high levels of noise and pollution. Although SUVs produce more carbon dioxide than motorcycles because of their much higher fuel consumption, motorcycles are 20 times more polluting with respect to carbon monoxide, unburned hydrocarbons, nitrogen oxides, and particulates (soot). Motorcycles might not be as "bad" as some of their riders make them appear, but they are truly dirty. Putting emissions controls on a motorcycle would significantly detract from performance and make the bike less satisfyingly loud. If experience is any guide, many bikers would simply disconnect the unwanted equipment.

The number of motorcycles on the roads is increasing, causing pressure to eliminate the loopholes in antipollution and antinoise laws that enable these bikes to disturb the peace. The costs of owner-ship will increase, which might help turn consumers to a new option: the electric motorcycle. The Vectrix Corporation of Middletown, Rhode Island, has begun selling a 440-pound zero-emissions bike capable of 62 mph, using nickel-hydride batteries (similar to those in the Toyota Prius hybrid) with an expected battery life of 10 years. The Vectrix e-motorcycle costs approximately $11,000. Brammo Motor-sports of Ashland, Oregon, has come up with a lighter, 275-pound bike, the Enertia, using lithium-ion batteries with an even longer expected lifetime, for about $12,000. Valence Technologies of Austin, Texas, makes the batteries.

The electric bikes now on the market are still a bit more expensive than comparable gasoline-powered bikes. But they have no tailpipe emissions, and operating costs (for electricity) are considerably lower than for conventional motorcycles. Even if the electric power is gen-erated by burning coal or natural gas, the electric version will be about twice as efficient (in life-cycle terms) as the gasoline-powered version.

The average middle-aged suburban commuter in the United States isn't likely to buy and use either a powerful gasoline-powered motorcycle or an $11,000 electric scooter to ride to work. But the costs of the e-scooter have already begun to drop radically, both because the lithium-ion battery technology is still rapidly improving and because China will soon sell much cheaper imports. Moreover, even if e-bikes are limited to 15 or 20mph, they will be able to use the bike paths or reserved lanes that many cities are building into their traffic plans for the coming decades. The falling cost of an e-bike can mean a falling cost for transportation energy service—a potentially significant stimulant to the urban economy of the transition years.

We expect that electric bikes and scooters will soon become very popular in Europe and some parts of the United States, providing at least a partial solution to the suburban-sprawl transport dilemma. Culturally, it will take a huge change for Americans to shift from big, powerful SUVs and cars to small, light vehicles—especially one-person vehicles with no doors or roofs. But it was a great cultural change to go from movie theaters to video games, or from snail-mail to e-mail. (Each of those shifts was only a partial replacement but represented a highly significant one from both a cultural and an economic standpoint.) And if just 20–30 percent of the short trips in cities shift to lightweight electric vehicles, fuel use and emissions will drop enough to help secure the transitional bridge.

And for All Who Still Need Cars...

Another big part of the motor-vehicle solution for both cities and suburbs will be car sharing. Switzerland first tried this idea in the late 1980s, mainly in connection with the Swiss railway system. It soon spread to Germany. It has been slower to catch on in the United States, but it appears to be gaining ground. A survey conducted by a University of California researcher in 2007 found 18 car-sharing programs in the United States, with membership growing steadily. At latest count, these programs had about 400,000 members using 7,000 cars.

The standard car-sharing business model is similar to the Velib bicycle-sharing model now operating in Paris, except that car-sharing sites now tend to be concentrated in central-city locations such as railway stations. Users pay an annual membership fee and rent cars by the hour at modest rates (but with no free period). The weakness of the system today is that users must reserve cars in advance and pick them up (and park them when returning) in specific locations. The cost of a four-hour use in the United States averages around $30. For multiday trips, rates are comparable to those of standard car rental. A number of car-sharing ventures have failed, but the shakeout is showing signs of helping to transform urban mobility. For a city dweller who needs a car only two or three hours a week, car sharing is financially attractive because the driver doesn't have to pay seven days' (168 hours') worth of parking, insurance, and depreciation for two or three hours of use.

The City of Philadelphia has contracted with car-sharing firms since 2004, eliminating 330 vehicles from the city fleet. Zipcar, the largest car-sharing firm, claims that each Zipcar takes 15 other vehicles off the road, each member drives 4,000 miles per year less than before joining the program, and each member saves an average of $435 per month by using the service. This is a true double-dividend and negative-cost solution, and it's available today with existing technology. New technology will make it even more competitive.

Like car-rental firms, car-sharing firms now use standard vehicles. However, because the mode of usage is different—a shared car might make several short trips each day—it will (like a taxi) accumulate mileage much faster than a conventional vehicle. Typically, a rental car or taxi is put out to pasture (sold second-hand) after two years of intensive usage and 120,000 miles on the odometer. At that point, it will be ready for its third set of tires, and will be showing distinct signs of mechanical wear in other areas. Yet many of the components of such a car are still nearly new. Such cars might be well suited to the car-sharing business, provided that the car-sharing firm carries out regular and rigorous maintenance. For the purposes of the energy-transition bridge, car sharing offers at least two advantages. First, by reducing the number of cars owned, it reduces the need for parking and driving space and helps set the stage for more compact urban designs (see the next chapter). This reduces the distances traveled by *all* modes—with commensurate reductions of fuel use and emissions. Second, by shortening the car's life span (in years), it can help get gasoline-burning cars off the road sooner and accelerate the transition to electric cars.

It might seem curious that we have gone this far into a chapter on vehicular contributions to the climate and energy bridge without any discussion of hybrids. With the unexpected success of the Toyota Prius and other gas–electric cars, hybrids have become the poster children of the green-tech revolution in America. They're a tribute to the vision (and business savvy) of Toyota and Honda, and a rebuke to the complacency of GM, Ford, and Chrysler, who were still scorning Al Gore's call for replacing the internal combustion engine. But hybrids might be more important for what they've done to change the culture than for their impact on fossil-fuel use, which is relatively small. As we noted earlier in this book, the typical gasoline-powered car gets about 1 percent payload efficiency, and a hybrid raises that to only 2–3 percent, at best. Hybrids, by themselves, don't provide a major girder for the transitional bridge. As part of a multifaceted change in the composition of urban mobility, they do.

Automakers will almost certainly build the next generation of plug-in electric cars mostly from aluminum and plastics, using advanced lithium-ion batteries. These cars will offer performance (range and acceleration) that's slightly inferior to that of conventional gasoline- or diesel-powered cars, but with zero tail-pipe emissions. They will be well suited to commuters and local errands, but not for long trips. They will offer low operating costs, but still at significantly higher prices. During the next two decades, manufacturing costs should decline significantly, thanks to technological improvements in the batteries and to economies of scale and experience. The spread of all-electric cars might be significantly accelerated if central cities offer preferential parking and plug-in facilities. But the largest gains will come in the long run from redesigning *cities*—and beginning that redesign now, to adapt to climate change—to sharply reduce the need for private motor vehicles altogether.

We can gain some limited insights by comparing the energy efficiencies and emissions of different kinds of vehicles per passenger mile. For example, we know that a conventional diesel bus emits only three-fourths as much carbon dioxide per passenger-mile as a car. But how the various modes are deployed is just as important as what modes they are, if not more so. In the Introduction to this book, we noted that the real key to the energy-transition bridge is not new energy supply, but new approaches to energy *management*. How the deployment of buses is managed makes a bigger difference in fuel use and emissions than the difference between buses and cars. With optimal integration of transport planning and urban design, instead of emitting 75 percent of what a car does per passenger-mile, the bus's carbon emissions can be cut to about 17 percent. We discuss how that can be done in the next chapter.

8

Preparing Cities for the Perfect Storm

We see two compelling reasons for taking present action to cope with distant, still hard-to-see threats. The first, which we have discussed throughout this book, is the need for *mitigation*—the near-term reductions of fossil-fuel combustion and carbon emissions needed to reduce the severity of future climate-change damages, as well as to find a more sustainable growth path and break free of imported–oil dependence. The second, which we focus on here, is the need for *adaptation*—preparation for changes that at this point we can no longer stop.

The consensus of climate science tells us that no matter how good we are at mitigation (and we'd better be very good at it), catastrophic extreme-weather events are headed our way. Katrina was only a warm-up. In organizing this book, we postponed this part of the discussion to maximize the effect of our argument that concerted action now and in the next few years will spell the difference between an economic recovery, even in the face of the coming catastrophes, and a relapse to conditions reminiscent of the 1930s or worse. But having said what we can about the possibility of cost-effective mitigation to get us through the next quarter-century, it is essential also—and simultaneously—to prepare for what we will encounter later, on the far side of the economic chasm. That "perfect storm"—the confluence of peak-oil aftermath, declining oil-based technologies, and climatic catastrophes—could be full-blown before we're across the transition bridge.

The first stage of adaptation to the long run must be a *part of* the bridge. Vulnerable areas worldwide can take a cue from California, where an interagency Climate Action Team in 2009 released the first of 40 reports on the impacts of projected sea-level rise during this century, and on the actions residents of specific areas need to take to avoid catastrophe. "Immediate action is needed," said Linda Adams, the Interagency Team's Secretary for Environmental Protection. "It will cost significantly less to combat climate change than it will to maintain a business-as-usual approach." The report noted that 260,000 Californians are already living in flood zones, and that a sea-level rise of 1.4 meters would increase the population at risk to 480,000. The number of miles of California roads and highways likely to be submerged would go from 1,900 to 3,500. And as we note a little later, it's not unrealistic to surmise that the coming rise in sea level—and consequent storm-surge devastation—could prove far worse than that.

Assuming smart mitigation strategy during the transition, the global impacts of what it is too late to prevent will still be so manifold that we can't begin to address them all here. They will affect global biodiversity, ecological stability, food supply, water resources, epidemics, geopolitics, and the survival of hundreds of millions of vulnerable people. In all these areas, if we do care about our grandchildren (whose prospects are less assured than the neoclassical economists tell us), and even the very survival of our species, we'll take long-term adaptive action sooner rather than later. And although these far-ranging impacts all have their own communities of experts racing to prepare defenses, they converge in one area: the approaching megathreat to many cities.

Although every city on Earth (and, for that matter, every town, village, and isolated farm house) will be affected, we focus here on American coastal and delta cities that lie in the paths of the most certain threats—rising sea level and increasing intensities of hurricanes, storm surges, and floods. Think Miami, Jacksonville, New York, Charleston, Tampa Bay, Galveston, and Houston—the usual suspects. Also think of the two largest U.S. navy bases, at Newport News, Virginia, and San Diego, California. Think inland cities on rivers that swell when the rain or the mountain snowmelt far upstream is unusually heavy: Sacramento, Cincinnati, Louisville, Memphis, and St. Louis. The danger reaches far into the heartland: In spring 1997, the Red River of the U.S. northern plains began to swell, and by the end of April, the water in the city of Grand Forks had risen to 56 feet above its normal level—the height of a six-story building. By that time, fires had raged through the upper stories of buildings where electric wires or gas lines had been ripped loose by the flood, and

50,000 people had to be evacuated. A Grand Forks man who had fought in World War II remarked that the downtown district reminded him of Dresden, Germany, after Allied bombers pounded it in 1945. In the Mississippi and Missouri river flood of 2008, Cedar Rapids, Iowa—a city where almost no one had bought flood insurance because floods were considered so improbable—found itself inundated.

It was never going to be just New Orleans.

And for New Orleans, the tragedy of Katrina is not necessarily just history. The assailant will likely be back—perhaps emboldened—sooner or later. Katrina veered a little and struck a glancing blow, but a future hurricane could strike head-on. And for any future storm surge, less wetland will buffer the impact. A half-century ago, the Mississippi River Delta basin included about 215,000 acres of land; but that land—which absorbs much of the impact of any hurricane—has been eroding into the Gulf of Mexico at a rate of 1,000–3,000 acres a year (2–9 acres *each day*), and less than half of that original buffer now remains.

More than half of the U.S. population lives in the 772 counties that are on the coasts. Demographers project that, by 2025, nearly 75 percent of Americans will be living near the coasts. Worldwide, nearly two-thirds of all humans—about 4 billion people—live within 90 miles of the coasts, hundreds of millions of them in cities that are 10 or 20 times more populous than New Orleans. Some of these cities could be destroyed right now if a Category 4 or 5 hurricane and storm surge struck head-on. However, as sea level rises, the surge's reach greatly increases. In places such as Bangladesh or the Mississippi Delta, an increase of 1 foot in the starting height of a surge might take the flood a mile farther inland.

Forecasts of future sea-level rise vary widely, mainly because of uncertainties about how fast the Antarctic and Greenland ice sheets will melt. (The melting of the Arctic Ocean ice won't raise sea level because floating ice has already displaced the water under it.) The IPCC consensus forecasts a rise of 11–88 centimeters (5–40 inches) by 2100, but the range of possibilities is considerably wider. The first IPCC report suggested that the Arctic Ocean could be ice-free by 2100; subsequent estimates brought that date forward, first to 2070, then to 2050. Now it looks as though the fabled Northwest Passage will be open by 2020, if not sooner. Good news for supertankers, but bad news for polar bears and a host of other species—and for us.

James Hansen, director of NASA's Goddard Space Center, is more pessimistic. Hansen, who first alerted the U.S. government to the climate threat in the early 1980s, notes that the IPCC forecast doesn't take into account the possibility of rapid Antarctic melting (a factor the IPCC scientists didn't feel able to predict); and if sea-level rise continues to accelerate, as it has since 1950, it could reach 5 meters by the end of this century—a time some of our grandchildren will live to see. Even more disturbing is the worst-case scenario Al Gore described: If the Greenland and West Antarctica ice sheets both collapse and melt, the rise would be as high as 12 meters.

How would this impact U.S. cities? If you've seen Gore's film, *An Inconvenient Truth*, you know what happens to New York City. Good-bye, Wall Street. But for the moment, set aside that biblical scenario and just consider the more conservative IPCC outlook (88 cm)—with maybe a cautious look at Hansen's concerns as well.

In New York, Columbia University and NASA scientists working for the U.S. Global Change Research Program have projected a sea-level rise of 4–35 inches by the 2080s, slightly lower than the IPCC's forecast. For readers who have some familiarity with New York, it's worth noting that two of the Columbia/NASA scientists, Cynthia Rosenzweig and Vivien Gomez, have calculated that if sea level rises just 18 inches, the surge from a (moderate) Category 3 hurricane would put the Rockaways, Coney Island, large parts of Brooklyn and Queens, parts of lower Manhattan, and eastern Staten Island under water. The entire subway system would be flooded. In another study, the program's scientists projected that a Category 3 hurricane on a direct-hit track would cause a surge of 25 feet at JFK airport, 24 feet at the Battery, and 21 feet at the entrance to the Lincoln Tunnel. The scientists didn't say what would happen if the sea-level rise was higher, or if the surge was from a Category 4 or 5 storm.

In Houston, at a 2008 forum on future hurricane threats, researchers confirmed that surges on the Texas coast had reached 17 feet above sea level during Hurricane Ike—enough to reach a depth of 5 feet in places that were 10 miles inland. With even minimal oceanic expansion between then and 2020 or 2030, or with a more intense hurricane, the surge would go farther and deeper. In California, the 2009 Interagency report projected that at about the same sea-level rise that the NASA scientists in New York assessed, the San Francisco and Oakland airports would be submerged, as would more than 330 hazardous-waste sites.

If Hansen's projection or Gore's worst-case scenario should prove realistic, we won't be making fine distinctions between lower and midtown Manhattan, or between Galveston County and Chambers County, nor will those who have lost their homes stand bravely before TV news cameras and vow to rebuild. Vast areas of the U.S. coasts will be permanently or frequently under water and probably uninhabitable, forcing rebuilding on higher ground. If the ice sheets melt completely, half of Florida will be gone. By then, if we and our children are smart, large parts of our coastal and riverside cities will have moved. And *most* cities and towns, whether in storm-path locations or not, will have undergone revolutionary changes in their management of energy.

Again, we're not in the forecasting business. We simply note that these scenarios cannot be absolutely ruled out and that the consequences would be truly catastrophic. Perhaps it would make sense to invest in some protections in advance. The cost would surely be much less than a post-catastrophe cleanup and reconstruction.

Urban Metabolism

Cities aren't fixed objects bracing themselves for a hit. Like individual organisms, they're continuously growing, regenerating, coping with threats, and adapting. And like the ecosystems all organisms depend on to exist, they process energy and materials, and dispose of wastes. How well they do these things will determine how well they cope with catastrophe.

To draw an analogy, the average untrained man or woman can run fast for perhaps half a mile, at most, before exhaustion—before the energy system fails. A highly trained "ultra" distance runner can go 100 miles or longer, with only a few brief (two- or three-minute) stops for water, nutrients or elimination. The difference is in the efficiency of the long-distance runner's energy system—it's a *different system* than what the untrained person uses, resulting in 200 times greater total output. It's nature's proof of what "sustainability" really means—a system that can go on for a long time by maintaining a balance between rates of input and output. The untrained runner uses energy anaerobically—unable to process enough oxygen for the task—and builds up waste products, such as lactic acid, too rapidly to dispose of. He (or she)[1] builds an "oxygen debt," very much like financial debt that is too large to pay off, and comes to a paralyzing halt. The long-distance runner, thanks to his training, operates in an aerobic mode, relying on his more efficient circulation to deliver oxygen and energy to the muscles as fast as they are used. By avoiding oxygen debt or glycogen depletion, he is able to keep going for many hours instead of a few minutes.

In a similar way, cities can achieve far greater energy efficiency—and, not incidentally, far greater capacity to withstand shocks. From the standpoint of industrial ecology, American cities (and most cities of the world) are out of shape. The analogy with circulation is particularly apt. We so often refer to the highways around urban areas as "arteries" that the term is no longer metaphorical, nor need it be. Clogged arteries endanger an individual's life, especially if the individual is suddenly stressed. In the same way, the capability of a city to withstand the shocks of climate change is greatly weakened by congestion. We've seen what happens when a hurricane strikes a city such as New Orleans and people can't get out—it's a form of urban thrombosis. If a coastal city is struck by a storm surge, bad traffic circulation and inadequate public transportation could be fatal to a large part of its population.

But the problem isn't just limited to times of emergency. Poor traffic circulation in cities results in overconsumption of fossil fuels,

---

[1] The progress in women's distance running serves as an instructive lesson in how rapidly society can increase its capacities if the incentives are strong. Before 1984, the Olympics included no running events for women longer than 1,500 meters because it was believed that women couldn't run longer than that distance (about a mile) without harm. Activists campaigned, the International Olympic Committee changed its policy, and in 1984, one of the activists, Joan Benoit, won the first women's Olympic Marathon in a time that would have won the men's race a few decades earlier. Today women regularly compete in 100-mile races and longer. In their energy management, cities can take a cue from Joan Benoit.

resulting in unnecessary emissions of carbon dioxide and other green-house gases:

- Stop-and-go driving (what if the blood in your arteries did that?) causes poor fuel economy. The very fact that we actually treat this circulatory irregularity as normal, by establishing separate fuel-efficiency standards for highway and urban driving, is a measure of just how dysfunctional our cities are.
- Heavy reliance on conventional gasoline-powered cars instead of greater reliance on bus rapid transit, electric cars, scooters, bicycles, and walking results in far higher per-capita energy use than necessary.
- Suburban sprawl results in higher per-capita distances traveled, whether for commuting, shopping, education, or recreation.

Those ills have been amply discussed elsewhere. We bring them up here mainly to emphasize that cities are continuously rebuilding and regenerating, even if they don't migrate away from their riverside or coastal origins. Over time, we can redesign their circulatory and metabolic systems. During the period of the energy-transition bridge, whole urban districts will need to undergo the first stages of redesign. Although the primary purpose of urban redesign will be to reduce energy use and emissions (to *mitigate*), the redesigning also provides greater opportunity to prepare for whatever future climate-change damages the mitigation doesn't ultimately avert (to *adapt*).

Mobility

Although a significant share of individual mobility in urban areas can shift from cars to bicycles and e-bikes, with significant reductions in energy use and emissions, another (overlapping) share can shift to public transit with similar benefits. Many Americans might be resistant to giving up cars, but as congestion worsens, this becomes an increasingly attractive option.[2] Millions of New Yorkers (and Parisians, Romans, Berliners, and Amsterdamers) either don't own cars or use

---

[2] A World Bank study found that passengers using a new bus rapid transit (BRT) line in Beijing cut a trip that averaged 60 minutes by car to an average of 37 minutes by BRT.

them only for weekend trips out of the city; for in-town mobility, they find the subway and bus systems more convenient. After all, you don't have to find parking for a subway car. Even in sprawling Los Angeles, public transit is catching on as both commuter trains and bus rapid transit (BRT) systems have induced growing numbers of commuters to gratefully stop driving to work on the I-10, Ventura Freeway, or other corridors of frustration and road rage.

For the transition-bridge strategy, new BRT systems warrant particular attention because they are faster and less expensive than rail systems. BRT got its original launch in the city of Curitiba, Brazil, where planners half a century ago projected that rapid growth in the number of people and cars would require massive expansion of the road system. As an alternative, they redesigned the city's growth so that instead of expanding haphazardly in all directions (as most cities do), it would develop along designated corridors served by BRT lines radiating from the center "hub."

Unlike conventional buses, BRT systems have some of the same features as subway systems: unimpeded movement without traffic lights or congestion, frequent service (on some of Curitiba's lines, a bus goes by as often as every $1\frac{1}{2}$ minutes), and quick passenger loading and unloading from door-level platforms. Today Curitiba has 2.2 million people, and 70 percent of them use BRT to commute to work. Compared with eight other Brazilian cities of similar size, Curitiba uses 30 percent less fuel per person.

Light rail systems take longer to build and cost more, but as with Curitiba's bus system, they can bring far greater order and efficiency to a city's metabolic system. Houston's MetroRail began operating in 2004 and was expected to have 45,000 riders per day by 2020. It proved far more popular than anticipated, reaching that target by 2007. The electric Lynx Blue Line in Charlotte, North Carolina, was expected to reach about 9,000 riders per day in its first year but instead was getting 16,000 by its ninth month in operation. It's not just in New York and San Francisco; all over the country, for millions of people who want to get from here to there without spiking blood pressure, the affair with the private car is over.

In comparisons of alternative transportation systems with regard to fuel use and carbon emissions, BRTs offer huge gains. A study of two new BRT systems in Los Angeles (along the Wilshire Boulevard and Ventura Boulevard corridors) found that they're saving 19,000 barrels of oil per year. The Washington, D.C.–based Breakthrough Technologies Institute measured carbon dioxide emissions per passenger in various urban transportation alternatives, with the following approximate results in grams per passenger mile:

Personal vehicle	400
Conventional 40-foot diesel bus	275
Light rail	210
BRT 60-foot hybrid	125
BRT 40-foot compressed natural gas	65

Not surprisingly, in countries that are parties to the Kyoto Protocol, BRT systems have become central to urban planning. Cities in Colombia, Chile, China, Laos, Panama, Peru, and Brazil are all establishing such systems. By 2009, 63 BRT systems were operating on six continents, and at least 93 more are in the works. Yet even those represent only a small fraction of the global potential both for reducing global energy consumption during the time of the transition bridge and for reducing risks of congestive failure in times of storm-driven evacuation.

Buildings

The main tool for making cities less vulnerable to rising sea levels, storm surges, and floods will be updated building standards—comparable to those that have been enacted in recent years to make buildings in Japan and California more able to withstand earthquakes.

The most immediate risk new standards must address—if seawalls or dikes are breached—is the damage or destruction of natural gas lines and electrical wiring, and the equipment used for cooking, refrigeration, lighting, and water purification. You might not expect that a primary danger of flooding is fire, but as we noted in our recollection of the Grand Forks flood of 1998, the scene after days of inundation was not just of buildings immersed in water, but of their upper stories in flames. And the danger is not only of fire, but also of electrocution, contamination of drinking water, and disease. Building standards—and architects—will need to move the electrical guts of the buildings in these areas well above ground level, preferably on or under the roof. And backup solar power (if not an independent micropower system) would provide an extra measure of protection if the grid itself were to fail.

We should pause here to distinguish between two different concerns: the capabilities of cities to withstand a climate disaster, and their capabilities to operate with high energy efficiency all the time. The latter is critical to the energy-transition bridge because it applies to *all* cities and towns, whether they're in an obvious path of harm or not, and because high efficiency mitigates the severity of future harm wherever it might occur. In an interconnected world, higher energy efficiency in mile-high Denver contributes to a safer future for hurricane-vulnerable Houston. We might need different building standards for electrical equipment in different disaster-risk zones, but we need high-performance materials and energy efficiency everywhere.

In recent years, we've seen a progression from the increasingly popular "low-energy" houses offered by such builders as Pulte Homes (which had built more than 15,000 Energy Star homes as of 2008) to "zero-energy" buildings. If solar photovoltaic (PV) is added to zero-energy design, the building can then be called "energy plus," meaning that it produces more energy than it uses, with the owner selling the surplus back to the grid. For owners who don't want off-grid systems that require storage batteries or fuel cells, the grid then serves as an unlimited storage device. And as plug-in electric cars become more common, the batteries in the cars will give the system as a whole more storage capacity for smoothing periods of high and low demand on the grid.

Although retrofits can be costly, designing low-energy or zero-energy buildings from the ground up, especially if whole tracts or neighborhoods are involved, offers another major opportunity for low-cost, high-return contributions to the energy-transition bridge. For example, in Guangzhou, China, a new 71-story office building, the Pearl River Tower, combines high energy-efficiency design with both solar- and wind-power generation to operate at zero net-energy consumption.

Ultimately, all cities and towns will incorporate engineering and design improvements, and retrofits will provide continuing gains in national energy efficiency and micropower generation. Many of the retrofits on older homes or buildings will be modest (replacing old single-pane windows, plugging heat leaks, adding insulation, and so on), but some will be dramatic. In San Jose, California, Integrated Design Associates remodeled a tilt-up concrete 1960s structure, and it reopened in 2007 as a so-called "z-squared" building—achieving both zero net energy and zero carbon emissions. It uses a range of readily available energy-conservation techniques, such as optimal use of sunlight to minimize electric lighting during the day, ground-source heat-pump cooling, advanced insulation, and low-e (double-pane) windows.

The largest potential for cutting energy use in buildings is in reducing the need for space heating. No new technological development is needed to accomplish that. The well-proven techniques of high insulation, low-e windows, and passive solar design make it work. The best proof is in the European Passive House project, which began a few years ago and has achieved energy-use reductions of 90–95 percent compared with existing (often very old) houses, and reductions of 50–65 percent compared with typical *new* houses. The first European passive house was built in Darmstadt, Germany, in 1990, and it used 90 percent less energy than a standard new house at that time. In Germany, annual energy use averages about 210 kilowatt-hours (KWh) per square meter for existing structures, 95KWh for typical new construction, and 20KWh for passive houses. By 2007, about 8,000 passive house dwellings had been built in Germany. That's only 1 percent of the total housing stock in Germany, but with much of the initial institutional resistance now relaxed, the pace is accelerating.

In Austria, where winters are fairly cold, the primary energy use is about 240KWh per square meter per year for an existing house and 130KWh per square meter for typical new construction, but it's just 20KWh for a passive house. The first passive house in Austria was built in 1995; three were built the next year; and by 2006, there were 1,600. With the markets now growing exponentially, Austria expected (before the 2008–2009 economic slump) to be building 50,000 units a year by 2015. Most other European countries are following a similar path. In Norway, where people have unsurpassed experience in coping with the cold, existing houses have a much lower energy use of about 100KWh, but the new passive houses cut that to just 10KWh per square meter per year.

In launching its Passive House project, the European Commission aimed to have all new-home construction meet the new standards by 2015. To overcome the predictable barriers of established "business as usual" markets and practices, the Commission set up a Promotion of European Passive Homes (PEP) campaign to provide easily accessible web-based information for all stakeholders—from architects and builders to city planners, financial institutions, national governments, homeowners, and renters. The transformation will dramatically reduce European energy demand over time, and it appears to be off to a fast start. In the United States, the first passive house (in Berkeley, California) wasn't completed until 2009.

In American cities where whole districts might need to be moved or rebuilt, designing from the ground up facilitates considerations that might not be possible with retrofits, such as orienting buildings and streets for optimal use of sunlight and more efficient use of living space. We can design the roofs of whole neighborhoods for solar energy collection, and the net energy use and emissions of those neighborhoods might go to zero. If city officials wake up to the importance of coordinating normal new-development planning with beginning the relocation of low-lying districts, they will achieve the *triple* benefits of minimizing future global warming, avoiding the storm-surge damage that future warming might bring, and providing an urgently needed new boost to economic recovery and growth.

Rooftop Revolution: The Solar PV Opportunity

The advanced home-building technologies of zero-energy, z-squared, or passive houses will provide dramatic energy savings in the long run, as older structures are eventually replaced. In the short term, however, the potential for these gains is limited because the more advanced technologies require new construction.

That slow pace of improvement might change as the costs of PV power[3] come down. As with both the decentralization of electric utilities and the redesign of cities, bringing PV to its full potential will take decades, but the first stage will be an important part of the transition bridge. PV costs are declining rapidly—enough to have triggered a boom in the stock prices of solar manufacturing companies, even as the overall stock market was staggering.

The Prometheus Institute for Sustainable Development estimated that up-front costs of rooftop PV panels dropped from $6 per watt in 2004 to $5 in 2009, and would fall still further by 2013. The Earth Policy Institute has projected an even faster drop. The leading U.S. thin-film PV manufacturer, First Solar Inc., an Ohio-based company that also operates plants in Malaysia and Germany, has developed an efficient manufacturing process for a thin-film cadmium-telluride PV cell that is cheaper to make than the older poly-silicon cells. First Solar has estimated that manufacturing costs could drop as low as $1 per watt in the near future—a cost that would be competitive with a coal-burning central power plant.

At least two possible clouds hang over this bright prospect: scarcities of critical materials and problems with toxic waste from discarded end-of-life solar panels. Tellurium is extremely scarce—the Earth's crust contains only 1 part per billion (ppb), compared with 37ppb for platinum, for example. Tellurium is a byproduct of copper refining, and scientists don't know of any ores from which it can be extracted. The First Solar process requires about 10 grams of tellurium (along with nearly 9 grams of cadmium) per square meter of thin film. That adds up to about 135 metric tons per gigawatt (GW)—which happens to be almost exactly equal to the world's entire non-U.S. output of the metal in 2007. If we used it all to make solar panels, their capacity would be about one-tenth of 1 percent of U.S. electricity output.

---

[3] The future importance of solar power is a consequence of a simple—but rarely acknowledged—fact: The sun each day provides about as much energy as humans use in a year. The problem is that most of the sunlight is too diffuse and in the wrong places for optimal industrial use. Photovoltaic cells can help overcome this problem.

Fortunately, other possibilities exist and research is going into high gear. Other thin films might come into play, and worldwide PV growth has continued at a rapid pace. Growth is fastest in countries where governments have provided strong incentives to help the industry get established. Germany, which instituted a "feed-in" tariff (requiring utilities to pay a premium price for power received from small producers), had 300,000 buildings with PV rooftops—57 percent of the world's installations—by 2007.

U.S. installations—spurred by a $2,000 tax credit against construction costs that the Energy Policy Act of 2005 provided, and California's "million solar roofs" program—jumped 83 percent in 2007. But the United States still has only 7 percent of global installations and lags far behind several other countries. On the manufacturing front, U.S. performance lags even further. Although worldwide PV manufacturing has been growing at 48 percent per year since 2002—reaching 3800MW in 2007—U.S. production ranks behind Japan, China, Taiwan, and Germany.

The $60 billion allocated for U.S. investment in renewable-energy projects by the American Recovery and Reinvestment Act in 2009 will help spur the lagging U.S. movement toward a more sustainable urban economy in the long-term future. But as we have stressed, the greater near-term need is for investments that will more quickly reduce the costs of energy services and carbon emissions. The shotgun Recovery Act provides little targeted support for energy recycling, CHP, utility decentralization, or other girders of the transitional bridge.

Armor versus Agility

A looming question for path-of-harm districts—one that hasn't exploded into public consciousness but likely will with the next Katrina-sized or larger climate catastrophe in a North American city—is whether devastated districts should rebuild or relocate. And for those who are willing to try to manage that event before it happens, a more immediate way to put the question is whether the most effective strategy is to "armor" urban waterfronts with stronger levees and sea walls, or to begin moving development back without waiting to see if the armor is strong enough to take a hit.

No single right answer exists for all vulnerable districts. Some cities have less flexibility than others. New York's Manhattan has nowhere to go, and its great concentration of skyscrapers, infrastructure, and capital investment is so immense that armoring is probably the only option even for a worst-case sea-level rise and storm surge. New York Congressman Anthony Weiner said in 2007 that New York's most underutilized asset is its waterfront, which suggests the possibility of synergistically combining sea-wall construction with post-oil infrastructure, such as a solar-PV rim around Manhattan Island combined with an around-the-rim e-bike and pedestrian path. But for less compact areas, massive walled-city solutions would likely be far more costly than orderly relocation. For New Orleans, the hope of long-term protection by levees might be delusional.

Historically, one of the first reactions of people who have had their houses destroyed by a hurricane or flood is to stand before a TV news camera and declare, "We will rebuild." In his book, *The Control of Nature*, John McPhee tells the story of a family who built their California dream house in the foothills of the San Gabriel Mountains, in a place that was directly in the path of two very predictable dangers— wildfires and torrential runoff. Periodically, when hard rain follows a fire that has denuded the steep hillsides in that region, the resulting saturation of the ground can trigger calamitous mudslides. One night, such a slide came down the canyon behind the family's home, smashed through the back wall, and quickly began burying and filling the house with mud. As the mud level rose, the family retreated to the upper floor and clung to a bed, which floated toward the ceiling. The family members lifted their faces to the diminishing airspace above the bed and said their good-byes, and then the mudflow stopped. It was a traumatically close call, and a brutal reminder to the family that in canyons such as theirs, mudslides happen repeatedly. Yet, asked what they would do, they said they'd rebuild in the same place.

That has been the common response of uprooted people after every natural disaster almost everywhere in the world, from Sumatra after the tsunami in 2004 to the Midwestern floods in 2008. New Orleans was no exception. But when the devastation extends to a whole coastal or floodplain community and the vow is to rebuild the waterfront with a stronger and higher sea wall, it's a little like the reaction of a medieval knight who's been brutally knocked to the ground in a joust with a much more powerful knight. Does it make sense to climb back on his horse for another encounter? It might show valor, but if the knight values his life, how smart is it? For an armored district that has been flattened, that probably depends on whether retreat is feasible. In the case of the Netherlands or New York City, there is really no place to go. So the Dutch have undertaken a massive engineering project to protect their country, and it's apparently working. But the hugely expensive Dutch solution will not be feasible in many other places—and might not work in a worst-case scenario of sea-level rise.

Among city planners and public officials, we have seen an awakening to the danger of climate change as a general phenomenon, but not so much as a threat of direct assault on their own bailiwicks. On February 16, 2005, the day the Kyoto Climate Accord finally went into effect (with the United States almost alone among major industrial nations in its refusal to sign), the U.S. Conference of Mayors issued a "U.S. Mayors Climate Protection Agreement" to advance the goals of the Kyoto treaty through local actions. The agreement committed participating cities to "strive to meet or beat the Kyoto Protocol targets in their own communities," through actions ranging from anti-sprawl land-use policies to urban forest restoration projects, and to "urge their state governments and the federal government to enact policies and programs to meet or beat the greenhouse gas–reduction target suggested for the United States in the Kyoto Protocol." The mayors of 710 cities signed the agreement.

It's politically easier to fight climate change as a general threat that is, to some degree, still an abstraction, than to tackle the very specific, costly, and intrusive task of beginning to move whole districts of one's own city out of harm's way. Beyond the hornet's nest of protest that even suggesting such relocation will stir up, a significant factor in its non-discussion might be a sense that such a project is simply too far outside the traditional functions and responsibilities—and financial capabilities—of local government. In Chapter 4, "The Invisible-Energy Revolution," we noted that one impediment to companies pursuing aggressive energy-efficiency measures has been the view of many managers that energy efficiency is not part of the "core business" of their companies. Similarly, for many public officials, anticipating and protecting against a possible future devastation of their cities might seem as outside their core business as preparing for military invasion.

A related impediment might be the reluctance of politicians (who have been sometimes characterized as more often followers than leaders) to express views that are outside the mainstream. Mark Twain wrote in 1905 that the more intelligent a man is, the more likely he is both to harbor unpopular opinion and not to utter it, as "the cost of utterance is too heavy; it can ruin a man in his business, it can lose him his friends; it can subject him to public insults and abuse...." It's fashionable now to advocate sustainable policies, but still too radical to suggest, on a day when the sky is blue and no hurricane is on the horizon, that it's time to begin an evacuation. If it's hard for neoclassical economists to recognize that something terribly disruptive has happened to their long-held expectations, it's also hard for public officials.

However, events will eventually drive a path-of-harm exodus, even if cities haven't prepared for it. The question is whether to prepare as an armored knight does or as an expert in jujitsu does. The choice to be made—whether to rely on heavily armored sea walls or to avoid heavy blows—will become a more publicly debated issue. It emerged as an incipient issue in 1993, when a summer-long flood inundated nine states of the Mississippi River basin, destroying 50,000 homes and 15 million acres of farmland. Putting heavy strain on the federal flood insurance program, the disaster triggered a serious policy debate about whether the best way to protect people is to strengthen and heighten flood walls or to move them off flood-prone land. The decision at the time was to move a lot of farms off the bottomland, but to defend heavily developed riverside cities such as Des Moines and St. Louis with higher levees and walls.

There was an exception, however, that might offer an instructive precedent. The town of Valmeyer, Illinois, about 30 miles south of St. Louis, was completely swept away by the flood, and it was fairly clear to everyone who witnessed it that if the town were rebuilt in the same location, it would sooner or later be swept away again. And as events turned out, it *would* have been destroyed again in summer 2008. But after the flood of 1993, the county's regional planning committee decided to move the entire town to a bluff 2 miles away and 400 feet higher. In 2008, the relocated residents were able to watch safely from their windows as the river raged past in the distance far below.

The Valmeyer move provided an opportunity not just to avert future devastation, but to make a more comprehensive preparation for the post-oil future. With assistance from state and federal agencies and the Working Group on Sustainable Redevelopment, the new Valmeyer was designed to be one of the most energy-efficient communities in the country. Its new Emergency Services Building, which houses the police and fire departments and some of the municipal offices, was built with state-of-the-art solar and energy-efficiency technologies, as were a new school and library, apartment complex, and senior center. New single-family homes were built with high insulation, low-e windows, low-flow showerheads, water-conserving toilets, and energy-efficient heating and cooling systems. Research was begun to assess the possibility of generating wind energy for the town.

Of course, there's a big difference between a 950-resident town such as Valmeyer and a real city. But we're not suggesting that whole cities need to move—only that their most low-lying or waterfront districts may. The new Valmeyer was built on a 500-acre tract. Most large American cities have at least that much land available on higher ground—even if only a few meters higher—in blighted districts that need redevelopment whether a catastrophic event is headed their way or not.

In redesigning for climate change, the precipitating motivation might be a hurricane, major flood, or near miss. But that epochal project will also serve as a stimulus to the resource-efficiency upgrading of entire urban regions. For example, sea-wall or relocation projects will create heightened attention to urban design possibilities such as retrofitting existing structures for rooftop solar-PV energy and better insulation. For those efforts, at least four key goals are paramount:

1. **Space-conserving development**—New higher-ground districts should be more compact than the low-lying development they replace. Benefits include shorter distances between home, work, shopping, and other areas, meaning less travel per person and a larger share of trips that people can reasonably take by e-bike, bicycle, or walking—resulting in less mileage per person and less energy consumption per mile. Less of the urban space

will be turned into pavement, meaning less dark-colored surface absorbing sunlight and radiating heat (urban "heat island" effect) that directly adds to global warming.

2. **Public transportation oriented**—In some U.S. cities, planners have needed to rationalize building metro lines partly on the grounds that they would revive blighted areas. For example, a proposed extension to the Washington, D.C. subway was approved because it would bring new vitality to a part of the city about a mile northeast of the White House that had become a crime-ridden area of abandoned buildings and empty lots. Similar rationales have been proposed for blighted areas of Los Angeles. With new high-ground developments planned from scratch, public transit (most likely bus rapid transit) can come first, with the new construction designed around it. Engineers can design car-sharing sites into the system instead of their being an urban-development afterthought. Similarly, engineers can plan a network of outlets for plug-in electric vehicles. The final result of such ground-up planning might be far more successful than when redesign requires an *ad-hoc*, tail-wagging-the-dog process.

3. **Living space oriented**—With public transit providing the main means of personal mobility and personal vehicles made secondary, we can achieve compactness without sacrificing living space. The space savings come from paved areas, not kitchens or living rooms. With car traffic subordinated, planners can provide even more outdoor living space (gardens, parks, and pedestrian spaces), along with cleaner air and a generally safer neighborhood environment. Networks of bicycle, walking, and running paths can augment public transportation, allowing carbohydrates instead of hydrocarbons to provide some of the mobility—resulting in far lower carbon emissions (just people and their dogs and cats exhaling) and a healthier urban population.

4. **Low-energy, low-emissions buildings**—One of the great silver linings of a prospective path-of-harm exodus is that a new design offers greater opportunity than retrofits for reducing energy use and emissions. As the European Passive House experience shows, old buildings have abysmally poor energy efficiency—and in European cities, where many buildings are

centuries-old stone structures, energy-efficiency gains have to come mainly from new construction. In the United States, with its larger share of wood-frame houses and higher rates of both demolition of old structures and new development to meet higher population growth, the benefits of enlightened new home construction can be even greater. It's not unreasonable to expect that within the period of the energy-transition bridge, we could upgrade a large share of American houses to zero-net-energy, zero-emissions structures in neighborhoods where the risks of flood damage, auto exhaust, power blackouts, and kids running into the street have all significantly abated.

9

The Water-Energy Connection

"Peak oil" has been a subject of welling concern—articulated in books ranging from academic to alarmist; in peak-oil blogs; in tense unpublicized meetings of oil company engineers and executives; and in damage-control advertising by oil companies aimed at reassuring us that there's still an "ocean of oil" to be found—at slightly higher prices. Technically, peak oil will be an epochal moment. In reality, it probably won't be recognized until after it has passed. What we need to be most concerned about is the years *after* that recognition. With human population expected to continue rising until midcentury, and the peaking of global oil production expected several decades before that, the gap between supply and demand will begin to open with staggering speed.

It's impossible to predict in detail what will happen in the aftermath. But we can be fairly sure that if we haven't crossed the energy-transition bridge by then (and we probably won't), world order will be severely tested. Beyond the widely held view that the Iraq War was largely about oil, we have already seen other signs of possible trouble: heavy Chinese investments in the Sudan; kidnappings of oil executives in the Nigerian Delta; the hijacking of an oil super-tanker off the coast of Somalia; brazen thefts from oil wells in Texas; and acrimonious disputes over future rights to drill for oil under the arctic ice, in residential neighborhoods, or in the Arctic National Wildlife Refuge.

We have seen striking parallels in the world's growing struggles over fresh water. Unlike oil, water won't get "used up," at least in the literal sense. But with fresh water, too, we will see a widening gap between supply and demand worldwide. In fact, in per-capita terms, we have already passed "peak water." The quantity of available *fresh* water on Earth is declining as aquifers are emptied but not recharged, and as mountain glaciers melt and drain into the salty oceans.

With water, as with oil, we see signs of incipient civil disorder: In India, farmers and a Coca-Cola bottler clash over access to a declining aquifer; in China, conflicts arise between farmers, industries, and urban users over water tables that are rapidly falling. In the North American West, cities are rapidly buying up water rights from farmers who need cash. California, Nevada, Colorado, Arizona, and neighboring Mexico are engaged in a long-running conflict over access to the water in the Colorado River, which is now pumped dry by the time it reaches its mouth. In recent years, just as the term "oil wars" began to appear with growing frequency, so did the term "water wars." A Google search at the time of this writing turned up 68 million references to "oil wars" and 137 million references to "water wars."

In public discourse and media, impending scarcities of oil and water have been treated as separate issues. But they are profoundly linked in two ways.

First, consider that on a biological level, water and energy are both essential to life—to the metabolism of every organism or community. Energy without water, whether from the sun or embodied in food or fuel, can't keep a single person, dog, or tree alive.

Second, at an industrial level, water is essential both to help extract useful forms of energy from raw materials and to employ the harnessed energy in the major industries of civilization: agriculture, manufacturing, construction—and the energy industry itself. As much water is used in harnessing energy (primarily in cooling electric power plants) as in any other human use, with the possible exception of agricultural irrigation. Industrial and municipal washing both require water. It's the closest thing to a universal solvent. It's the primary carrier for fertilizers, paints, detergents, dyes, inks, acids, and alkalis used in industry. It's the carrier for human wastes. It's also the primary carrier for heat.

Useful work doesn't have to be provided by burning oil products, such as gasoline or kerosene. But as long as internal combustion engines or steam turbines perform much of the work, the water-oil link will become increasingly troublesome. We are entering an era when conflicts over access to either fresh water or petroleum (or gas) can lead to war. Shortages of either can threaten supplies of the other.

In a larger sense than the person who first said it probably meant, oil and water don't mix. To grasp just how dangerous this marriage of convenience has become, it's helpful to look more closely at where the water is in today's world—and where it intersects our present energy system.

The planet's fresh water is mostly locked up in the form of ice, primarily in Antarctica and Greenland, plus some mountain glaciers of which the greatest volumes are in the Himalayas and the Andes. Of the remainder that is liquid, the largest amounts are in a few lakes, such as the Great Lakes in North America, the lakes of the Rift Valley in Africa, and the lakes of northern Canada and Siberia. Another quarter is believed to be permanently circulating in the Amazon basin and its tributaries. Worldwide, the seasonal melting of snow and ice in the mountains feeds the great rivers, such as the St. Lawrence, Mississippi–Missouri, Columbia, and Colorado in the United States, and a score of others around the globe. The rivers are one of the two primary sources of fresh water for human uses. In California, for example, snowmelt from the Sierra Nevada feeds the American and Sacramento rivers, which form one of the primary sources of water for the state's 37 million people. The other primary source in much of the world is rainfall.

Both snowmelt and rain are typically seasonal. Spring runoff on most continents, and the monsoon months in South Asia, create a great water surplus followed by months of drought. So, water has to be stored when it's plentiful, whether in aquifers, reservoirs, or cisterns. Interestingly, the problem of water storage is quite similar to that of solar or wind-energy storage. The demand for energy is 24/7, whether or not the sun is shining or the wind is blowing. One of the major tasks in ramping up renewable-energy industries will be to greatly improve energy storage and distribution technology—and infrastructure. Apart from fossil-fuel inventories, our largest capacity for stored energy is currently also our largest stored-water system: the reservoirs above hydroelectric dams.

But hydroelectric power provides only a small share of the U.S. energy supply, and it's not likely to grow. We're draining many reservoirs faster than nature can replenish them. For example, the water level in Lake Meade, below the Boulder Dam in Nevada, has fallen by 100 feet in the past two decades—partly because of the growth of nearby Las Vegas and partly to satisfy the demands of Southern California on the other side of the Mojave Desert. The need to develop efficient, small-scale storage devices for electric power is increasingly critical—one of the reasons we have stressed the need for a transition bridge from fossil fuels to sporadically available renewables such as wind and solar energy. Similarly, with billions of people living in places where long dry spells follow rainy seasons or monsoons, water storage is increasingly critical.

When the human population was smaller and more closely clustered near year-round water sources, people had less need for water storage—and some of the storage was in the form of storable food grains. Rivers played key roles in the rise of civilization, both as continuously flowing sources of drinking or cooking water and as irrigation for food crops. The water of the Nile, flowing through a region of harsh desert, supported the agricultural development of ancient Egypt. The surplus output from crops planted in the silt left from the annual flooding of the Nile subsidized the creation of cities and central governments—and their sustenance through dry seasons. Silt deposited by the annual flooding of the Tigris and Euphrates rivers enabled the earliest inhabitants of the Middle East to make the same transition. Much the same story seems to have played out in the valley of the Indus, the Ganges–Brahmaputra, and the valleys of the Yangtze and Yellow rivers of China.

However it began, civilization now depends on agriculture, and agriculture depends on fresh water. With today's global population a thousand times as large as it was at the time of the early Sumerian or Egyptian civilizations, rivers and canals alone no longer suffice to serve agricultural needs. In some fortunate countries and regions, such as France, rainfall is sufficient (irrigation is needed only in peak summer months, if at all), and replenishable sources can provide the fresh water. But the bulk of the world's cereal-grain crops—the main crops for human consumption either as food or as feed for livestock, poultry, and fish farms—now depends on irrigation from ground water. Unfortunately, the aquifers that provide that water are being depleted almost everywhere.

Most irrigation is now done by drilling wells. Some aquifers are naturally replenished by rainfall and runoff, but most are not. Well-feeding rotary sprayers now thickly dot the U.S. grain belt west of the Mississippi and east of the Rocky Mountains. From above, the land looks like green polka dots on a brown background. Electric pumps extract the water, mostly from the giant Ogallala Aquifer—a long underground lake of "fossil" water left over from the melting of the last glaciers of the Ice Age 10,000 years ago. The annual runoff from winter snows and summer rains in the Rocky Mountains provides some replenishment, but not nearly enough. The Ogallala is being depleted at an estimated 12 cubic kilometers per year—more than the annual flow of the Colorado River. This rate of depletion cannot continue much longer. In parts of Texas, Oklahoma, and Kansas, water tables are dropping by as much as 100 feet per year, and much of the Ogallala could run dry in as little as 25 years. As the water level drops, the energy required for pumping from ever-deeper levels increases. But without irrigation, grain cultivation in the southern Great Plains becomes impossible and the land will revert to native grasses or mesquite. If we don't undertake water-conservation measures soon, the Dust Bowl of the 1930s could return in spades.

We're also diverting river waters into irrigation canals (where much of the water is lost to evaporation), leaving the rivers and their downstream valleys dry. The Rio Grande and the Colorado rivers are both examples of this tendency, with disastrous consequences for northern Mexico. The Aral Sea in Uzbekistan has shrunk to a tiny fraction of its former self, thanks to water diversions to irrigate cotton fields that were planted in the 1950s and 1960s. The tremendous evaporative losses and other diversions from the Aswan Dam on the upper Nile have sharply cut water flow to the lower valley from 32 billion cubic meters a year before the dam to only 2 billion today. The Yellow River in China, like the Colorado, doesn't reach the sea in most years. China also is diverting water from the headwaters of the Mekong, leaving less for Thailand, Laos, and Vietnam. Turkey is taking more water from the upstream sources of the Tigris and Euphrates, leaving Syria and Iraq with less. Not surprisingly, tensions are rising.

That brings us back to the water–energy nexus. As in our discussion of urban metabolism, it's helpful to consider the relationship between human and industrial metabolism. Cities and civilizations were originally developed to support human physical needs for year-round food, water, shelter, and waste disposal, so it's not just coincidence that urban resource management aims to provide the same basic services that the human body requires.

For individual humans, the energy source is food, which is made up largely of carbon-hydrogen-oxygen compounds from plants (carbohydrates). Similarly, the dominant industrial energy source today consists of carbon-hydrogen compounds formed from plants that grew hundreds of millions of years ago (hydrocarbons). Nearly all food other than ocean fish now comes from agriculture, and today's agriculture consumes vast amounts of water. It takes roughly a thousand tons of water to produce 1 ton of grain (wheat, corn, or rice). According to Lester Brown of the Earth Policy Institute, the amount of water required to produce food for the average American is about 2,000 liters (more than 500 gallons) per day.

The United States is fortunate, because rainfall supplies a lot of that water, and only one-fifth of the country's grain has to come from irrigated land. Other countries aren't so fortunate. In India, three-fifths of the food comes from irrigated fields; in China, it's four-fifths. Those two countries combined have about eight times the population of the United States, and their farms' demand for water is rapidly depleting their natural aquifers. For example, water tables have been falling at a rate of 20 feet per year in the Indian state of Gujarat. When a fossil aquifer is pumped dry, rain won't replenish it for centuries. In northeastern Iran, the water table fell by about 3 meters a year during the 1990s, but the wells are now dry because the aquifer is empty. The people who previously lived above those aquifers have become "water refugees." Although only a fraction of the agricultural land is irrigated in the United States, that fraction accounts for a large share of the nation's wheat, sorghum, and corn production—the basic energy source for both our human metabolism and that of the cattle, chickens, or fish we raise for food.

The international implications are stark. The ability of China and India (not to mention that of Pakistan, Iran, and a hundred other countries) to feed themselves is now under threat. Declining groundwater availability translates directly into reduced grain production in the few countries of the world that have historically had grain surpluses. Thanks to increased use of fertilizer, China had a surplus for a few years before 2004. But in that year, China had to import 7 million tons of grain because of water scarcity in the northern part of the country that is normally irrigated by the Yellow River.

Indian grain output might also soon begin to drop. Countries with increasing water shortages and growing populations will need to import grain to cover the deficits. Grain trade can be regarded as a form of water trade. But just as the number of petroleum-exporting countries is declining, so is the number of grain-exporting countries. We may be approaching a time when scores of countries that don't have enough water to produce all their own grain will need to buy grain from elsewhere, but the exporters won't have enough to sell.

In the United States, as we noted earlier, the electric power industry uses a large share of fresh water to cool power plants. In 2000, power plants used 136 billion gallons per day, which amounts to about 450 gallons of water per person per day, just for our electricity. By now, that amount has probably increased. Much of this water is returned to the rivers or lakes from which it's drawn, so not all is "consumed" (made unavailable for other uses for a long time). However, the newer plants now use closed-circuit cooling instead of the older "once-through" systems, which means that the amount of water tied up by power plants is increasing as new capacity comes online. In the older, once-through plants, waste heat is continuously carried back into the rivers or lakes, where it can do ecological damage instead of being harnessed for heating homes or buildings through a combined heat and power (CHP) plant as discussed in Chapter 2, "Recapturing Lost Energy."

The first engineering response to the water problem is usually to build giant aqueducts to transport water from places of surplus to places where local supplies can't support the population. In China, water from the huge Three Gorges hydroelectric project on the Yangtze River will be pumped to the northern part of the country, including Beijing, at a rate of 44.8 billion cubic meters (44.8 cubic kilometers) of water per year. But the construction and operating cost of the south-to-north diversion project will be enormous. The water pumping requirements will likely consume a significant amount of the electric power generated at the Three Gorges Dam. Indian engineers are undoubtedly contemplating similar projects to divert water from the Himalayas to the south, especially to Madras and the state of Tamil Nadu.

California has been the prototype for this sort of response. The entire southern part of the state, including parts of the Central Valley, the Imperial Valley, and the 18 million people in the urban belt from Santa Barbara to San Diego, depend on water from the northern part of the state or from the Colorado River. Much of the water from the north is pumped over mountains as high as 2,000 feet, and the electric power to operate the pumps for interbasin transfer alone now consumes 6.9 percent of the electric power used in the state. (The U.S. national figure is about 3 percent.) Water and sewer systems, as a whole, consume 19 percent of the electric power used in the state and 33 percent of all non-power-plant natural gas used in the state.

It takes more than 5,400 kilowatt-hours (KWh) of electric power to pump 1 acre-foot (43,560 cubic feet) of water, or 1KWh for every bathtub-full, from the Sacramento–San Joaquin delta to Cherry Valley, the most distant destination in southern California. By comparison, the electricity requirement for ocean water desalination is 3,800–4,400KWh per acre-foot. The electric power needed to deliver fresh water from the north to the southern end of the California system is actually greater than the electric power that would be needed to desalinate an equivalent amount of ocean water. Other options are much less energy intensive: Groundwater purification by reverse osmosis (depending on location) averages about 1,000KWh per acre-foot. For groundwater pumping alone, the energy requirement—based on California data, and probably also applicable to the Ogalalla Aquifer—averages 2,250KWh per acre-foot per year.

From an energy perspective, the best source of water for the Los Angeles Basin would be recycling the wastewater from sewage treatment. For irrigation purposes (without reverse osmosis), the electric power required for sewage treatment is only 400–500KWh per acre-foot; with reverse osmosis (needed for household consumption), it's about 1,300KWh per acre-foot. That's still a lot less than is needed to pump water from northern California or the Colorado River. The reuse solution might not appeal to many people, but it's exactly what happens in nature. In Europe, this process is already a necessity.

Fresh water arriving in London from the upper reaches of the Thames is recycled eight times before reaching the sea.

According to the 2005 California Energy Commission's Integrated Energy Policy Report, "The state can meet energy and demand-reduction goals comparable to those already planned by the state's investor-owned energy utilities ... by simply recognizing the value of the energy saved for each unit of water saved ... . Initial assessment indicates that this benefit could be realized at less than half the cost to electric ratepayers of traditional energy-efficiency measures." This process is another potential double dividend, but regulations limiting the investment possibilities of regulated electric utilities currently block it.

Although the water–energy conflict looms largest in agriculture, it's rapidly spreading to other realms. In Colorado, Wyoming, and Utah, water interests and oil companies are locked in battle over the future of oil-shale exploration, which could have a staggering impact on the future not only of the mountain West, but also of much of southern California. Shell Oil and other companies are agog at the theoretical potential of the oil shale there; some estimates suggest that several times as much oil might be locked in the rock under those states as there is in liquid form under Saudi Arabia.

Unfortunately (or perhaps fortunately), no viable technology currently exists for commercially extracting the oil from the rock. As we have seen with innumerable cases of past technological development, it takes many years to develop a major new industrial technology, build the plants and infrastructure needed to put it to work, and bring it up to a scale where the costs are competitive. As with "clean coal" and other diamond-in-the-sky hopes for carbon, oil shale has no chance of contributing to the U.S. energy supply during the period of the transitional bridge. And it's quite possible that by the time oil shale does eventually prove feasible and affordable (if it ever does), we will have finally crossed the bridge to the carbon-free future and it will be obsolete.

Processing oil shale will apparently require prodigious quantities of water (as steam), which would mean taking large amounts from the Colorado River, which is already overtapped. Just how much water would the processing require? No one knows for sure, because the technology isn't that far along. Some experts estimate that it will take ten barrels of water to produce one barrel of oil from shale. "There are estimates that oil shale could use all the remaining water in the upper Colorado River basin," said Denver Water Board commissioner Susan Daggett in a 2008 interview. In October 2008, the U.S. Congress allowed a moratorium on oil-shale development of federal land to expire, triggering an oil-shale rush similar to the Coal Rush described in Chapter 5, "The Future of Electric Power." The oil companies even received a chunk of the initial $700 billion financial-sector bailout to assist them in their quest. That money would have been better spent on water conservation.

Water-Energy Goals

Like impacts of climate change on cities, the increasing scarcity of water and resulting human conflict will be a long-term problem requiring solutions that take decades to fully achieve. The distances that water will need to be pumped will continue to increase, and the energy costs will continue to rise. But as with the relocation of low-lying urban districts to higher ground, the long lead time required for remedy is all the more reason to get started as soon as possible. As with urban redesign, the first phase of water-management reform must constitute a girder of the energy-transition bridge.

The importance of this nexus might not be fully visible yet, but we may be on the threshold of a "positive feedback loop" (vicious cycle) that will cause the energy cost of water to skyrocket as global temperature rises. We are already witnessing some of the worsening impacts:

- **Desertification**—During the past several decades, large areas of the world that were previously arable, or at least marginal, have turned to desert. The main causes include overgrazing and deforestation. For example, 50 percent of the cropland in Kazakhstan has been abandoned because of desertification since 1980. In Iran, advancing sand storms have reportedly buried more than 100 villages. In Africa, the Sahara Desert is advancing into Ghana and Nigeria by more than a thousand square miles per year. A U.N. University study projects that if current trends continue in Africa, the continent might be able to feed only 25 percent of its population by 2025. And in northern China, an estimated 1 million acres is turning to desert each year. With climate change, desertification will likely accelerate. As farmland shrinks, demand for irrigation rises, and water tables fall further.

- **Forest-fire cycling**—Global warming is making forests drier, which increases the frequency and intensity of wildfires. Fighting fires is becoming a year-round industry in much of the U.S. West, consuming a growing share of stored water as thousands of fire trucks and aircraft scoop water from reservoirs to drop on flames. The fires also send more carbon dioxide (not to mention heat) into the atmosphere, contributing further to the warming—and, ultimately, to further drying of the forests.

- **Salt contamination of fresh water in coastal areas**—As the sea level rises, ocean salt increasingly invades coastal groundwater and estuaries. Aquifers serving coastal cities are ruined, and rivers reaching the coast are turned brackish farther upstream.

All these impacts will increase the need for energy-intensive water transport or pumping—from deeper wells as tables drop, from reservoirs as air tankers and fire trucks battle wildfires, and from ever farther-inland sources as salinity contaminates local supplies in coastal areas. In California, a 2005 study by Matt Trask of the California Energy Commission found that the amount of energy needed for pumping, transporting, and distributing water was 11,953 billion KWh per year, which was nearly equal to the 12,482 billion KWh required for the heating, cooling, or pressurization of water in its innumerable end uses.

The goal of water-management policy cannot realistically be to increase the supply. It must be to reduce per-capita water use in parallel with decreasing fossil-fuel use. The good news, in a sense, is that water waste, like energy waste, is enormous in the United States—with great potential for conservation. Aside from the manifold other benefits of water conservation (water security, food security, lowered political tensions, and better public health), more efficient water use will reduce energy consumption for pumping and for such energy-intensive uses as the cooling of nuclear or fossil-fuel power plants. Perhaps the largest near-term opportunity for easing the water–energy tug-of-war, until more of our electric power generation shifts to CHP, is suggested by a 2007 report by Argonne National Laboratory for the U.S. Department of Energy, "Use of Reclaimed

Water for Power-Plant Cooling." The study notes that more than 50 U.S. power plants are now using reclaimed water—most of them in California, Florida, or Texas, which have all experienced growing concerns about freshwater supply.

An abundant body of research has identified ways to reduce per-capita use by consumers, from the familiar water-saving showerheads and toilets to low-water landscaping and xerescaping. However, the biggest consumption is in agriculture, and the biggest potential for conserving water—and water-pumping energy—is drip irrigation. This technology uses 30–70 percent less water than conventional sprinkler systems. In the United States, 80 percent of all water use is for agriculture, but only 7 percent of U.S. irrigated land is drip-irrigated. (According to the U.S. Geological Survey, 47 percent is flooded and 46 percent is sprayed.) One of the most egregious wastes of water in the United States is the production of corn ethanol (remember Chapter 6, "Liquid Fuels: The Hard Reality"), which from irrigation to processing consumes 10,000 gallons of water for each gallon of fuel produced.

As water use becomes more efficient, we might barely notice the difference, because the benefit will simply offset what would otherwise be an alarming increase in the energy costs of pumping, power-plant cooling, and desalinization of ocean water. To make more visible gains, water managers might take a cue from industries that are profitably harnessing waste-energy streams. Recycling wastewater will very likely provide the most reliable—and cheapest—new water supply in the decades ahead.

10 ——————————

Policy Priorities

Policy discussions in America are too often a turnoff, except when they're driven by religious or political ideology or "not-in-my-backyard" conflicts—in which cases they result in hugely distorted views of what really matters to the survival and well-being of our civilization. With few exceptions, the truly large issues, such as human population growth, climate change, or accelerating biodiversity loss, are earnestly discussed in academic forums and journals, but barely touched in mainstream media.[1] Perhaps the best mainstream (or nearly mainstream) public-policy conversation in recent years has been that of public TV moderator Bill Moyers, whose guests have explored such issues as the power of pharmaceutical corporations in shaping American wants and values, and the pervasive conflict between science and ideology, which we have touched on in this book. Yet in 2008, Moyers's program, *Now with Bill Moyers*, attracted fewer than 3 million viewers on average, compared with the 30 million who were listening to right-wing pundit Rush Limbaugh on radio, or the more than 97 million who watched the Superbowl football game during Moyers's final year with *Now*.

Unfortunately, the policy issues we now face in building an energy-transition bridge are as critical as any the country—or humanity—has faced. The threats that now loom are on the same scale as those of World War II or the nuclear arms race of the Cold War, and—we believe—far more serious than international terrorism. The conundrum we face is that, from the standpoint of public

---

[1] Climate change is an emerging exception, which might signal a breakthrough in public awareness at this make-or-break moment in human history.

arousal and mobilization, the idea of recycling waste heat or eliminating the smoke from combustion of dirty fuels just doesn't grip the
public imagination the way that Al Gore's vision of a zero-carbon,
renewable-energy future does. Yet the former is an essential bridge
to the latter. And if any more drama is needed, we can promise that
what's at stake with this bridge will generate at least as much sus-
pense as any issue that has ever gripped us. The policies that will
enable us to successfully span the transition years are fairly obscure
matters of legislative and executive management, but how we manage them will affect how hundreds of millions of people will live or
die in the coming decades.

In Chapter 3, "Engineering an Economic Bridge," we briefly
listed the eight main girders that we believe will be needed to support
this transitional bridge. Chapters 2 and 4–9 described those girders
and the functions they will serve in reducing fossil fuel use, cutting
global-warming emissions, and boosting economic growth by making
energy services cheaper. (Chapter 1, "An American Awakening,"
probed the economic and scientific arguments.) We now return to
the list of main girders, focusing on the actions that can best facilitate
their construction. Because we've entered a time in which families,
governments, and businesses are in survival mode and scrambling for
solutions, many remedies are being proposed or attempted, and we
don't pretend to know which few of them might prove unexpectedly
promising. However, experience suggests that some of them will
waste critical time and resources. To keep our focus on the actions in
which we have high confidence, we emphasize two core principles
that warrant brief explanation before we describe the policies that
they dictate.

Core Principle 1: Reversing Sisyphus

The first requirement of effective energy-bridge policy is that it
provide incentives to increase both energy efficiency and economic
prosperity. *Most existing incentives work in the opposite direction.*
The crux of the problem with these existing incentives is that coal and
oil companies have profit incentives to sell more coal and oil, no

matter what their TV commercials proclaim about their embrace of alternatives.[2] Similarly, electric utilities have incentives to make investments that add to their rate base (and, incidentally, sell more electricity). Home builders have incentives to sell bigger houses (which require more heating and cooling), car makers have incentives to sell bigger and more powerful cars (which use more gasoline), and so on.

Regional interests add to the distortions. The U.S. Senate has a disproportionately heavy representation of rural populations. That also means over-representation of rural industrial interests (especially mining and agribusiness). We can trace the corn-ethanol subsidies to that distortion. Similarly, the high "discount" rates that some economists advocate count present benefits as worth much more than future ones. The executives at Archer Daniels Midland or Monsanto corporations now have vastly more representation in the making of U.S. energy policy than do the people who will be living in San Diego or New York City in 2050.

If we could change the incentive system so that energy companies aren't selling a tangible *commodity*, such as oil or gas, but a *final service* (heating, cooling, or—better still—physical comfort), their behavior would change. Responsible executives would still ask, how can our company operate most profitably? But the answer would change: by producing the most service for the least cost—by using the *least* possible fuel energy. Instead of the consumer having the difficult job of figuring out how to reduce ownership costs of highly complex products, the people who have expertise in making those products would assume that responsibility. Shifting corporate roles from selling energy-intensive products to selling the services those products provide also shifts the incentives from using more energy to using less. It might take a long time to make such a fundamental shift in corporate business models. But understanding how the existing

---

[2] For anyone who has been impressed by the expression of social responsibility those commercials convey, it's worth noting that, until recently, those same companies were smoothly reassuring us that global warming wasn't a problem. Only when it became clear that the public was no longer buying that line did the energy companies began their blitz of "green" advertising. But behind the scenes, they continue to lobby for the entrenched policies that work against the bridging to large-scale renewable energy.

incentives have actually widened the chasm we face can help clarify the shape and structure of the transitional bridge.

Core Principle 2: Simplifying the Regulatory Patchwork

The second core principle is based on the hard reality that "externalities" (pollutants and other side effects that no one takes responsibility for) are pervasive in the modern economy because markets don't exist, and can't exist, for all environmental services—no matter how important those services are. A market can't exist for something that nobody owns, or that anybody can take or use. No one owns the air, so no market for fresh air exists. A man might defend his property with a shotgun, but to do so, he also needs a fence and a "No trespassing" sign. This isn't possible for the air above his property, or for the birds that fly over it.

Similarly, no market exists for most of the air and water pollutants that our world is generating in ever-mounting quantities. With limited exceptions, such as scrap metal or paper, no market exists for most wastes because nobody needs or wants to buy them. No one is offering to buy the greenhouse gases we're putting into the atmosphere. In the real world, the prices of most wastes are negative, or would be negative if a real market existed. The polluter would need to pay someone to take the wastes. Too often, that someone then dumps the stuff into a river, a sewer, or the nearest unguarded woods.

The wastes and pollutants keep accumulating. But in practice, how can any private business justify protecting something that is entirely in the public domain? We can't put a fence around a piece of the world's climate or the hydrological system (rain, rivers, and aquifers) and run it as a private business. The climate knows no boundaries. And on the purchaser's side, how many people would agree to pay for something such as sunlight or air that they'll receive whether or not they pay for it? That wouldn't be a market; it would be a tithe.

Because markets can't (or don't) work to protect the "commons"—the natural recycling and filtering of fresh water, the biodiversity of the oceans or rainforests, or the stability of climate—the response of governments has been a plethora of what economists call

"second-best solutions." Examples include the voluminous regulations of air and water quality, food safety, drug safety, worker safety,
driver safety, child safety, and handicapped-pedestrian safety; regulations on what is too toxic to recycle or what can't be transported
across state lines; air traffic and marine traffic rules; antilitter rules
and zoning laws; bans on lead in gasoline and on DDT; antismoking
laws and narcotics laws; motor vehicle fuel-efficiency (CAFE) standards; and so on. Although such second-best solutions mean well
(and have protected millions of people's lives and livelihoods), they
have become an ever larger patchwork of Band-Aids upon Band-
Aids. The result is a form of governance that is increasingly difficult
for ordinary citizens to monitor or even comprehend. And the rules
and regulations often end up creating even worse problems than the
ones they were originally designed to solve.

It's inherently dangerous for the public to lump together regulations that might be critical to the survival of civilization with the kinds
of relatively trivial regulations that have caused many people to become
cynical or angry about government "interference" in general. If public
perceptions lump together mandates affecting U.S. economic recovery
and protection against climate-change catastrophes with laws requiring
every exterior door of a home to have a stoop at least 36 inches long, a
critically needed sense of perspective—and urgency—might be lost.

Seeing this, it should not be hard for those on the political left to
understand, if not necessarily share, the vehemence of those on the
right who have called for less government. On the other hand, we
have passed the point where anyone but a recluse can fail to recognize that government has a more critical role to play now than ever.
Weak government can be as great a danger to human liberty and well-
being as a dictatorship, because it lets the pirates, polluters, and
muggers—and Madoffs—run amok. Conservatives, too, now recognize that *lack* of regulation has its downside. The financial catastrophe
of 2008–2009 was a sobering illustration of the risks. But the question
remains: How should we design regulations to help and not hinder?

The intellectual debate has been framed as a choice between
direct regulation by a government or international agency and indi-
rect regulation via "getting the prices right" by means of a market.
Until the 1990s, the main focus of environmentalists was on direct

regulation and the mechanisms for rule making and enforcement. In the 1970s, the focus was mainly on bans (of asbestos, DDT, tetraethyl lead in gasoline, ozone layer–destroying chlorofluorocarbons (CFCs), and so on) or on industry-specific or even process-specific emissions limits. The latter applied to sulfur dioxide, nitrogen oxides, and microparticulates from combustion processes that could not be banned. However, a gradual realization emerged over the years that direct regulation, especially as it became more specific about the details, could be economically inefficient. It also tended to create cumbersome administrative bureaucracies.

Since the 1990s, a growing understanding has emerged that other approaches might be more efficient, at least in some cases. The classic example (still largely confined to academic papers and theoretical models) has been the pollution tax, as exemplified by the carbon tax. Industry hates it, but economists approve of the idea, in principle. But "small government" conservatives object that bureaucracies would need to set the tax and enforce its collection, and that the government—having collected the tax—would spend the money.

The emerging consensus today seems to be twofold. First, where direct regulation seems unavoidable, the trick is to set broad environmental and energy-performance standards instead of specifying emissions limits or technological solutions. Second, where the tax approach makes sense in theory, the bureaucratic barrier could be avoided by creating a new "article of commerce," such as tradable carbon-emissions permits, using government powers to set up synthetic markets for those permits. A pioneering example is the now-controversial "cap and trade" system, which the European Union first implemented in 2004. We hasten to acknowledge that in such markets, as the European example illustrated, the design details are important. The first few years of the European system are widely regarded as a failure, resulting in windfall profits for some electric utilities (in particular) and almost no progress in reducing emissions. After some revisions, the European system of tradable permits seems to be working a little better, but only a little.

The second core principle is to systematically replace the patchwork of second-best solutions left over from earlier times with a simpler and more streamlined regulatory and tax framework that rewards energy (exergy) efficiency, encourages productive work, and minimizes harm at the same time. For example, the policies we propose for the energy bridge—and for the years beyond—would decrease the cost of energy services (useful work), thereby boosting employment and income by shifting taxes from payrolls and personal income to carbon emissions. Simultaneously, that tax shift would discourage the selling or burning of fossil fuels by increasing the cost of doing so. That restructuring would shift some of the burden of a troubled economy from those who are trying to make a living to those who have been making a killing.

Policy Priorities for the Energy Bridge

In this section, we list the main girders of the energy-transition bridge (which we originally introduced in Chapter 3) and the policies—guided by the two core principles—that we believe can best facilitate them.

1. Encouraging Recycling of Waste-Energy Streams

The good news is that thousands of businesses don't need any government help to begin profiting—and performing a great public service—by capturing waste-energy streams and converting them to electric power for their own use or to sell back to the utility. This is a win–win situation if the surplus electric power can be sold at a price that is attractive to both parties. The bad news is that, in most cases, this will reduce the electric-power utility's sales and, correspondingly, reduce its profits. Utilities lobby hard to prevent changes to their present profitable-monopoly situation. It will take political courage to oppose such large and powerful companies, which are also major contributors to political campaigns. Would-be energy recyclers *will* need government help—not money, but changes in outmoded laws—if they want to sell surplus power to someone directly across the street instead of selling it back to the utility at a low price that the utility decides. If a waste-energy recycler can sell to anyone who's in the market for electricity, it has far more incentive to generate solar, wind, recycled, or CHP power than if it's allowed only one potential customer who dictates the price. Conservatives say they believe in competitive free markets, so they should be in favor of this change.

The electric utilities will correctly point out that they have built and maintained the "grid" of power lines and transformers that now delivers electricity to all consumers at (relatively) affordable prices. They will go on to incorrectly argue (we've already heard them practicing this speech) that introducing local competition will destroy the grid. That scare tactic is unfounded. As we noted in Chapter 5, "The Future of Electric Power," the main functions of the grid—reliability and availability on a 24/7 basis—won't change. Business lost to local self-generation or CHP will be replaced by new business, especially from the drivers of new electric vehicles. The grid will continue to serve large-scale and long-distance needs. Yes, the utilities will need to do some re-engineering to compete with local producers exploiting otherwise-wasted energy. But they have more than enough advantages of scale and experience to compete successfully.

We think that by enabling more energy recycling, the system will be able to accommodate increased demand from plug-in electric or hybrid cars, which will soon appear in large numbers, *without* building new, high-cost central power plants. By avoiding the construction of new plants, they will avoid adding carbon emissions to the national total, while also making the system less likely to crash as a result of extreme-weather events, fires, or sabotage.

> **Policy need #1:** Rewrite the now-toothless Public Utilities Regulatory Policy Act (PURPA) in a way that no longer allows state utility commissions to avoid promoting free-market competition. The law should provide incentives to renewable-energy producers in the local sale and distribution of electricity. PURPA was originally supposed to *encourage* such competition, but it no longer works. It is stymied by gargantuan loopholes and by the fact that some states ignore it. The rewritten law should include at least three basic changes: (1) Eliminate the "avoided cost" requirement that lets a facility enter the market only if it can sell electricity to the grid at a price that saves the utility the cost of building new capacity; (2) allow competing facilities to sell power to the grid at higher prices than the utility's price if they generate power that has zero-carbon emissions and other pollution-reduction advantages over the utility's power mix; and (3) require that every utility purchase emissions-free power from anyone who can deliver it,

at a price no less than the utility's own retail price in that location (and higher than that price if emissions free, as noted previously). The law should direct every state to enforce this requirement. With these changes, thousands more industrial plants will have incentives to generate electric power from high-temperature waste heat or pressure drops.

Policy need #2: Abolish all state laws that prohibit nonutility companies from crossing roads with wires or selling electricity directly to other consumers. If a plant's own electricity needs aren't enough to warrant the capital cost of installing an energy-recycling facility, adding the capability to sell excess power to a nearby user will sometimes make the difference between failing and being able to stay in business and provide jobs. And when that happens, the amount of coal or gas that the utility would have used to generate that power will be eliminated from the nation's fuel use. The bottom line is that greenhouse gas (GHG) emissions will decrease.

Also see girder #2, below.

2. Ramping Up Combined Heat and Power (CHP)

As noted in Chapter 2, "Recapturing Lost Energy," decentralizing electric-power production by local cogeneration will enable the heat energy that distant "central" power plants are currently discarding to be used to heat homes, apartment buildings, office buildings, and shopping centers. This can ultimately reduce the amount of fossil fuel those aging plants burn by more than half. Measures to bring central power plants into the twenty-first century, as discussed for girder #5, will help. But a mandatory change in the competitive power market will produce the most direct boost:

Policy need: Pass legislation requiring electric power–distribution companies to purchase an increasing *percentage* of power from decentralized CHP, decentralized rooftop photovoltaic (PV) panels, privately owned wind turbines, or other small private power sources. The advantage of such a mandate is that the market will determine the price paid to decentralized producers (with the caveat that it be no less than that paid to utilities, as discussed for girder #1), instead of a legislatively set "feed in" tariff such as currently exists in several U.S. states and other countries. As with waste energy-stream recycling, the amount of coal or gas that the utility would have burned to produce the power it buys in this way will be eliminated from the nation's fuel use. This will help kick-start the micropower revolution. By cutting the cost of energy services (reducing fuel use per kilowatt), micropower will stimulate economic growth and reduce carbon emissions.

3. Ramping Up Energy Efficiency in Buildings and Industrial Plants

Here, as with waste-stream recycling, achieving the large potential that remains untapped requires an innovative business model *and* government encouragement through tax and insurance treatment. We know of many cases in which savings will pay for the investment within a few years. The main problem, apart from the need to provide better information to the public, is that, for most homeowners and many small and medium-sized businesses, the rate of return needed to induce them to make an investment in energy conservation (compared to other business needs or current consumption) has been much higher than the low rate of return on bank deposits or mutual funds.

We can address this problem by creating a new kind of energy service company (ESC). The ESC, a reputable company able to borrow money at a fairly low rate, offers to undertake and finance the changes needed to substantially improve a home's or business's energy efficiency—and to assume responsibility for paying the client's utility bills for a contract period. For the duration of the contract, the client will pay the ESC the same amount per month that it was paying the utility for the previous 12 months, suitably adjusted upward if there is any increased demand for service.[3] However, the ESC will be paying the utility substantially less than that, because of the lower consumption (from the increased efficiency) per unit of service. The

---

[3] The contract would probably need to specify a maximum number of kilowatt-hours over which the homeowner would need to pay for any excess used, so that the owner wouldn't have any incentive to use excess electricity that would cancel out the efficiency gains.

ESC will then be realizing a monthly net revenue for this account. During the life of the contract, the monthly difference between the building owner's payment and the ESC's payment will be the ESC's profit. After that, the difference will be the consumer's saving. Variants of the contract could enable the building owner to begin saving sooner, or enable the ESC to take profits for longer, depending on the size of the efficiency gain, the available interest rates, and so on.

To achieve their full potential, ESCs will initially require government support to provide information and to underwrite performance guarantees, at least for the first few years. Investments in ESCs should be as safe as savings in a savings account, but with a much higher return. Making investments in accredited ESCs tax free, like municipal bonds, could attract venture capital.

An important tool for achieving economy-wide emissions reduction—by establishing strong incentives to improve energy efficiency in industry—would be a national (if not global) carbon "cap and trade" system. One version of such a system—killed in the U.S. Senate in 2008—was a subject of heated political battle throughout the later Bush years. The administratively cumbersome nature of the proposed plan contributed to its demise,[4] as did conservative beliefs—widely promoted by industry lobbyists—that mandatory emissions mitigation would reduce economic growth and cost the country heavily. We think that controversy is misconceived and probably won't remain a barrier much longer—first, because it is clearly possible to construct large parts of the energy bridge at negative cost

---

[4] The cap-and-trade system that was initiated in Europe a few years ago, which was similar to the one that U.S. Senators Warner and Lieberman later proposed, assigned carbon-emission limits to each company based on industrial sector. It then allocated most carbon credits free to existing companies based on a "grandfather" principle. Only a fraction of the credits were to be purchased at auction. By limiting the total number of credits to be issued, demand for credits was expected to exceed the supply, resulting in a positive market price. However, the European system gave away too many free credits, resulting in a low market price and a correspondingly low incentive to introduce real emissions-abatement projects. The large coal-burning German utilities—notably RWE—counted the credits they had received for free as an implicit cost and passed on the costs to their customers, resulting in windfall profits estimated at ™5 billion for RWE alone. RWE disputes the number, but the resulting scandal has reflected badly on the cap-and-trade idea itself.

or low cost, and second, because there's a way to do carbon trading that involves far less accounting and bureaucracy than the opponents of such plans have feared.

Policy need #1: Establish a federal program to regulate and underwrite energy service companies (ESCs) that provide small businesses and homeowners with start-up investments in energy efficiency or renewable-energy improvements.

Policy need #2: Design and set up a much simpler carbon cap-and-trade system than those that have been proposed or attempted so far. The objective of carbon trading is to reform the established system of energy pricing, which has allowed free dumping of greenhouse gases into the global commons. Attempts to establish carbon markets in the United States have stumbled, partly because systems such as the European system or the cap-and-trade bill rejected by the U.S. Senate in 2008 would attempt to specify maximum carbon-emission targets for each industry sector, enabling companies to obtain the necessary credits by creating "offsets." The trouble with offsets is that it's too easy to get credit for an investment that would have been made anyway and that causes administrative and enforcement difficulties.

A system that requires carbon permits *only* for the producers and importers of fossil hydrocarbon fuels or forest products—the coal, oil, gas, and timber companies—will be more effective. Permits would not be allocated directly to firms, whether by grandfathering or auction. Businesses that extract or import fossil fuels, wood, or nonfood agricultural products (to produce ethanol or biodiesel) would be required to purchase carbon permits on the open market in amounts determined by the carbon content of their inputs. The buyers would be mainly the big primary energy producers (coal, oil, and gas companies), and they would add the cost of the permits to the cost of their products. The added costs to the primary energy companies would be passed on to customers, affecting all downstream users like a tax. But the downstream users (such as manufacturing companies that use coal-generated power, or airlines buying fuel) wouldn't need to buy permits themselves, because the cost of their emissions would already be added to the cost of their carbon-intensive inputs.

The key is to allocate the credits annually or monthly *only to individual taxpayers.* The system would operate like an extended version of Social Security, or like the system of frequent-flyer miles. Taxpayer recipients could either sell the permits immediately through a computerized market or save them in hopes of getting a higher price in the future. To prevent hoarding, the permits would need to have expiration dates. (The government might need to regulate the creation of funds or carbon-based securities for speculative purposes.) The impact would be similar to that of the so-called "negative income tax" advocated by the conservative Nobel Laureate Milton Friedman many years ago.

The plan we outline here would vastly reduce the enforcement difficulty and administrative cost of the cap-and-trade schemes over which U.S. legislators have wrangled, because only the first tier (fossil-fuel-energy and bio-fuel or timber companies) would be directly affected. Market regulators would need to monitor the carbon transactions and offsets of just a handful of companies instead of hundreds of thousands of firms in many different industries. Those hundreds of thousands of firms would have a relatively simple path—on a level playing field—to making energy-efficiency improvements. Equally important, consumers of energy-intensive goods and services would have an economic incentive to conserve.

Finally, the money paid for these carbon-emissions permits wouldn't go to the government (except as the income to the individual sellers is taxed) and wouldn't be available for other government-spending projects. This feature should please conservatives who oppose government spending in principle. Meanwhile, allocating permits to individuals would also constitute an income transfer from higher-income (higher-consumption) groups to lower-income (lower-consumption) groups. This should please those on the left who want to see more equitable sharing in the benefits of a reviving American economy.

4. Continuing Efficiency Gains in Consumer End Uses

Here we address such free-standing consumer purchases as cars, power boats, gas-powered leaf blowers, portable room heaters, TVs, computers, and other energy-consuming products. In this domain, as in industrial energy efficiency, abundant negative-cost opportunities need little more than better dissemination of information to become self-sustaining. The growing popularity of compact fluorescent lights is an example. Policy changes—such as Germany's decision to prohibit the sale of incandescent light bulbs—can greatly accelerate progress. Two of these changes are among the fundamental reforms that our core principle #2 suggested—the need to replace counterproductive "second-best solutions" with remedies that don't have to fight uphill battles against incentives that work the opposite way.

> **Policy need #1:** Extend and strengthen motor vehicle gas mileage standards (corporate average fuel economy, or CAFE standards) to include *all* vehicles, including trucks, on a rising schedule—for example, leading to 45 miles per gallon (mpg) for all cars (which the Toyota Prius already achieves) by 2030, and 60 mpg by 2050. This is feasible, even with the (roughly) ten-year lead time it takes to bring a new design into mass production.[5] Trucks or SUVs that have been classified as "light trucks" shouldn't be excluded.[6] Similarly, CAFE standards should be established for airplanes.
>
> **Policy need #2:** Eliminate federal subsidies to oil companies.
>
> **Policy need #3:** Introduce an "extended producer responsibility" (EPR) law requiring companies that sell complex manufactured products (such as cars, TVs, computers, or printers) to take their products back at the end of their useful life for remanufacturing or recycling. This legislation should also incorporate measures to facilitate long-term leasing of durable goods instead of outright sale, because it would provide the

---

[5] Roughly five years to design and test the new body, engine, and drive train and the machines to produce them, and another five years to create the supply chain, build the machines and factory, and hire and train the workers.

[6] The light-trucks exception was put into law on the pretense that *of course* people don't buy that new Hummer or Lincoln Navigator to show muscle in the driveway or to take the kids to Disney World, but to do hard pickup-truck work such as hauling firewood or hay for the cows.

lessors (and manufacturers) strong incentives to promote operating efficiency and carry out good maintenance.

EPR is a radical idea for Americans (although it's close to adoption in Europe), partly because it is inherently difficult to implement and partly because, in some basic ways, it seems contrary to the whole notion of "ownership." The implementation problems include deciding and enforcing responsibility for the various components of the final product, such as the tires, batteries, and electrical systems of cars. Should the original manufacturer be responsible for components (such as brakes, alternators, or lights) that were replaced during service by components that other suppliers made? Sorting out these problems and organizing the logistics—actually "reverse logistics"—for collecting and returning end-of-life vehicles (or TVs, PCs, or refrigerators) to their makers will take a few years, but it will end a big part of the regulatory patchwork covering the myriad problems of solid-waste disposal.

A well-tested and effective way to encourage returns is to use a monetary deposit, which ensures that a product never loses all its value. A 25¢ deposit (or less) will virtually guarantee the return of a bottle or can. A $1 deposit will guarantee the return of any small battery. A $5 deposit will bring back any used printer cartridge. A $25 deposit will bring back virtually any portable electrical appliance or electronic device, such as a cellphone, PC, or TV. A $100 deposit would probably be enough to ensure that old cars are not abandoned. What works for cans or bottles will work for tires, cars, or anything in between. And responsible citizens will get their deposits back.

The capability of deposits to reduce waste and improve materials and energy efficiency has been known for years but has been only rarely and marginally employed. The manufacturers and dealers generally oppose return deposits. Dealers don't want to collect them, for space reasons, among others. Manufacturers don't want their old products back, even if they are remanufacturable (which is usually not the case).

Another part of the equation is the "junk" factor. Buy a car, and it's yours. If it wears out and you go to get a new one but the dealer won't take it as a trade-in, you might leave it on your property (if it's rural) or in an alley or empty lot, and let it rust. One of the authors of this book can look out his window at a neighbor who has 12 vehicles in his yard, only one of which is in use. Up the street, in another side yard, is a circa-1965 Volkswagen with a 9-foot tree growing through it. The United States might have 20 million to 50 million such derelict cars, trucks, and tractors.[7] In the past decade or two, they've been joined by countless TVs, washing machines, computers, and even such objects as abandoned medical devices containing dangerous radioactive materials. Scrap-metal dealers eventually pick up some of the discards, and others go to landfills. But in the United States, massive amounts of junk are left in places where they contaminate the water, soil, or air, or pose hazards to children.

A basic return-deposit system coupled with EPR (or "take back") policy would address this problem in a sweeping way, for the simple reason that a corporation making a complex product under such a law brings into play two complementary incentives. The consumer has an incentive to return the object to its place of purchase (or another designated site) instead of dumping it. The manufacturer has an incentive to reduce its final disposal costs by utilizing every possible recycling option (for various metals, plastics, and chemical residues) *and* has the requisite technical knowledge and dismantling capability (which the consumer doesn't).

How does such a policy help increase the energy efficiency of the product during its lifetime? One way is by funneling scrap materials more completely into recycling. For example, it takes far more energy to make steel from ore than from scrap. Recycled aluminum and copper save even more. EPR will provide a

---

[7] The United States has more than 250 million registered vehicles, but the number of derelict vehicles is unknown. However, the data from a few cities and states might be indicative of the volume. The New York City sanitation department picked up 146,880 abandoned cars in 1989. With increasing scrap-metal recycling, that number dropped to 9,200 in 2006. Michigan removed 92,000 abandoned vehicles in a recent year. However, those were just the cars left on public streets. The majority of derelict vehicles may be on private property, and not easily reached by recyclers.

huge boost to the use of "urban mines" (continuous recycling of depreciated or demolished buildings, bridges, highways, vehicles, industrial equipment, and consumer products) to reduce reliance on more environmentally disruptive extractive industries. Because it takes less energy to do urban mining than conventional mining or logging, it also means less fuel use and lower carbon emissions.

5. Decentralizing Electricity Production

The initial steps of waste-stream energy recycling and CHP (#1 and #2) provide a big part of this girder. Those steps—most important, rewriting the PURPA laws—establish a means of providing energy-saving competition for the big utilities at the local level. But the biggest part of the decentralization effort must tackle the utilities themselves.

> **Policy need:** Update the Clean Air Act to eliminate the loopholes that have enabled utilities to keep old power plants operating long past their time. The utilities' motive for keeping the old plants smoking is to avoid expenditures that would otherwise be required to upgrade the plants with modern pollution-control technology. Under the second Bush administration, federal appointees zealously guarded these loopholes, exempting power plants from Clean Air Act requirements to reduce carbon dioxide emissions, and promising to "grandfather" coal-burning plants that commenced construction before the new climate-mitigation rules took effect. This triggered a "Coal Rush" that threatens to further harm American air quality. In November 2008, the EPA's Environmental Review Board ruled that the EPA has no valid reason for failing to limit carbon emissions from coal-fired plants. But in his last days in office, President Bush bypassed Congress with a plethora of "executive orders" attempting to override the regulations. Clearly, power-plant loopholes will continue to be a political football until an airtight update of the Clean Air Act is passed.

Also: see girders #1 and #2.

6. Finding Alternative Ways to Provide an Energy Service

A recurring theme in this book is that what you're really seeking when you make a purchase at a Wal-mart store or Toyota dealer is not a product, but the service performed by that product. Unfortunately, what you are offered, in most cases, is an object made from plastic, glass, or metal that might provide the service you want. But the product isn't the service itself. You don't really go to the real estate office because you've always wanted to own a giant wooden box—you go because you want a home.[8] You probably don't dream of owning 2 tons of steel, plastic, and rubber; you want mobility and comfort. We don't satisfy our needs and fuel the economy by buying houses, cars, and computers—we buy comfort and security, mobility, and entertainment.

Sometimes those needs can be met in other ways that don't require buying or even using those particular products. In many cases, it's possible to reduce energy use by using a different, less energy-intensive kind of product to perform the same (or comparable) service. Classic examples include telecommuting and Internet shopping as substitutes for driving cars to workplaces or stores, or using buses or bicycles instead of private cars for short urban trips when conditions permit, or watching movies on cable TV instead of going to the Cineplex. About 40 million Americans now telecommute at least part of the time, and Internet shopping has become a fast-growing alternative to driving to a mall. A Nielson study in 2008 found that more than 85 percent of the world's online population has used the Internet to make purchases, "increasing the market for online shopping by 40 percent in the past two years." In principle, online shopping will also reduce energy consumption by enabling shoppers to stay home instead of going to the mall.

> **Policy need:** To encourage mobility substitution that sharply reduces energy demand, make the allocation of federal highway funds to states rigorously contingent on the creation of

---

[8] Rentals, whether of apartments, cars, or tools, are forms of services that shift some responsibility for efficiency from an individual consumer to a business that has more expertise in efficiency calculations. But rentals have only limited benefits in this respect. Even a rental apartment, which provides a form of housing service (you don't need to fix the roof or replace the hot water heater), doesn't satisfy the needs of its occupant until it's furnished with a variety of other products, from TVs to beds, which are typically owned by the renter.

bicycle paths, bus rapid transit (BRT) systems, recharging plugs for electric vehicles (EVs), and other facilities that help urban dwellers minimize the use of private automobiles. The legislation might allow a city with a dense central business area (with a daytime population greater than a hundred thousand) to obtain carbon credits by creating reserved lanes for high-occupancy vehicles and BRTs, bicycle lanes, and special parking reserved for EVs and car sharing. Surtaxes on large cars and congestion charges on automobile commuters can fund these programs. The legislation should also create a national "mobility bank," financed partly by congestion charges, to promote the construction of these alternatives to conventional automotive traffic.

Also: see energy service companies under girder #3.

7. Redesigning Cities

In the real world of the next half-century, the collision of urban planning and climate change will likely be complex and chaotic, almost beyond imagination. In the United States alone, it will entail literally millions of city council meetings, feasibility studies, and policy debates among public officials, civil society organizations, civil engineers, and federal agencies such as the U.S. Department of Homeland Security, Federal Emergency Management Agency (FEMA), and the Army Corps of Engineers or their successors, as well as media, public utilities and a spectrum of industries (real estate, construction, banking, and road-building, among others).

We don't pretend to be clairvoyant, and we claim no ability to simplify the complexities enough to propose a comprehensive or all-purpose plan of urban action in a few pages. One of the questions we can't address is the extent to which ice melt, sea-level rise, or upstream deforestation will increase the risks of catastrophic floods or storm surges for particular cities, and what particular measures will be needed to protect those cities in the coming years. We *can* address the prospect that such measures will be needed to a significant degree in a number of vulnerable American metro areas, and in hundreds of smaller communities. As we discussed in Chapter 8, "Preparing Cities for the Perfect Storm," what happened to New Orleans in 2005 was probably only the beginning. But getting people to either build adequate sea walls and dikes or move will be a tremendous challenge. Redevelopment should be on ground that is high enough and far enough inland to protect against a worst-case IPCC forecast for twenty-first-century ice melt, sea-level rise, and storm intensities.

Policy need #1: Launch a national campaign following the model of the European Passive House project, with a national-security priority comparable to that of the Manhattan project, to mandate that *all* new U.S. residential construction achieve better than a 90 percent reduction in energy use. A builder's average home-energy standard, similar to the corporate average fuel-efficiency (CAFE) standards required of auto manufacturers, can inexpensively implement the campaign. The government should-n't grant exceptions for McMansions classified as "light hotels."

Policy need #2: In each vulnerable coastal region, instead of waiting for an emergency no one planned for, states should establish long-term plans for evacuating barrier islands and beach communities and quickly begin phased execution of the plans. By providing incentives to move and disincentives to remain, the first stage can be voluntary. (See policy needs #3 and #4.)

Policy need #3: Pass legislation to ensure that increased risks to life and property are quantified to the extent possible and that they are determined on a local (block-by-block) basis, not averaged over a large city or region. Publish flood and storm-related risks, and ensure that local property tax rates and flood- and storm-insurance rates reflect the costs of protection and emergency response. Phase out subsidies for flood insurance and storm insurance, which are becoming increasingly unaffordable. Consider limited subsidies or tax rebates to encourage relocation from vulnerable locations.

Policy need #4: Encourage vulnerable cities and counties to acquire land on higher ground, adjacent to or within the existing urban area, for use in a phased relocation of low-lying neighborhoods. Rezone the most vulnerable districts, to ban further development in high-risk paths of harm, and establish a fund to compensate property owners who move voluntarily. Pay for the fund by establishing a surtax on new construction on virgin land, with exemptions for high-ground development meeting the needs of the planned relocation. In the event of major flood or storm damage (as in New Orleans), use eminent domain to take over areas not suitable for rebuilding and convert them to parkland. dunes, wetlands, and so on.

Policy need #5: Revise building codes for path-of-harm zones from which populations aren't actually relocated, to minimize destruction of gas lines or electric wiring and equipment in the event of storm surges or floods. Although it's hard to imagine that any large towns or cities will be completely uprooted and rebuilt elsewhere in this century, we can anticipate that substantial sections of cities now lying within a few feet of sea level will need to be relocated to higher ground within the same region, as we discussed in Chapter 8. In the meantime, upgrading building codes for those properties that remain in place can provide life-saving protections against the dangers of flood- or storm-caused electrical fires, loss of power, spread of sewage or toxic substances, contamination of drinking water, and consequent loss of normal protections against typhoid or cholera.

Also: see girder #6.

8. Linking Water Management to Energy Management

As we discussed in Chapter 9, "The Water-Energy Connection," the era of post-peak oil will also be a time of post-peak fresh water. Although the quantity of fresh water on Earth cannot increase, world population is expected to continue growing by more than 70 million people a year for years to come, and the global availability of fresh water per capita will fall steadily. The ongoing depletion of many aquifers will exacerbate that trend. In the United States, the most critical example is the Ogallala Aquifer, which extends from Nebraska to West Texas and irrigates some 20 percent of the U.S. grain farmlands. Rainfall doesn't fully replenish many "fossil" aquifers, including the Ogallala, and they're being rapidly depleted. The depletion will only accelerate as global warming increases the toll of drought. The great Dust Bowl of the 1930s, like Hurricane Katrina and the 2008 Midwest floods, might turn out to have been only an early warning.

During the time of the energy-transition bridge, the actions needed to reduce fossil fuel use will need to include actions to reduce water use. In agriculture, industry, and domestic use, water scarcity is becoming a growing crisis in its own right. In each of these areas, rising water scarcity and rising energy costs become mutually exacerbating. Because we can't increase the overall supply of fresh water, the actions that have the best chance of breaking the cycle are those that sharply reduce per-capita water consumption without reducing the quality of the services water provides.

Reducing water use is an effective way of reducing electric-power consumption for pumping, which accounts for 3 percent of the national total and as much as 7 percent in California.

> **Policy need #1:** Eliminate "farm" subsidies for corn ethanol, which consumes 10,000 gallons of water for each gallon of ethanol produced, and use the money to ramp up drip irrigation and other low-water strategies for food crops. Irrigation is the biggest user of water in the United States (and the world), and drip irrigation offers the largest single opportunity to reduce per-capita water consumption. For example, reducing water consumption in the Central Valley of California will help alleviate the dilemma of declining snowmelt in the Sierras, which is causing impending shortages of water sent down the California Aqueduct to the southern part of the state. Reducing the demand for long-distance pumping will also significantly cut electric power use.

> **Policy need #2:** Establish a U.S. Water Management Agency in the Department of the Interior to promote national and state programs to ramp up drip irrigation and waste-water recycling, use reclaimed water for electric power plant cooling, replace lawns with low-water landscaping, and make continued progress on water-use efficiency in industries, institutions, and homes.

Energizing the Future: Looking Ahead to Looking Back

Governments of the past are most remembered and appreciated now for actions they took that reached far beyond their own times. Although the American government of the first years after the Revolutionary War was very busy managing disputes over land ownership, taxes, voting rights, slavery, and the establishment of a national bank, what we most appreciate it for today is the Constitution. For all its shortcomings, that primary document, together with the first ten amendments (the Bill of Rights), has continued to provide a steadying hand more than two centuries after its authors have been gone.

There is an evolutionary explanation for this continuing appreciation: Human development is based on the transgenerational accumulation of knowledge and wisdom. We don't need to start from scratch to prove that flying machines are possible. We also don't need to repeat the mistakes of the Roman Empire, the Easter Islanders, or Mao Tse-Tung. Thanks to that societal learning, we can incorporate knowledge from past discovery into our institutional memory and build on it. The current governments of the world (including the U.S. government) have plenty of work to do, whether filling potholes, stabilizing the financial system, or catching would-be terrorists. But they'll be most remembered, assuming that institutional memory continues and our species continues to evolve, for how they responded to the perfect storm of linked crises—energy, economic, and environmental—that have converged in the first decades of the twenty-first century.

11

Implications for Business Management

From the first pages of this book we have maintained that, for the energy-transition bridge to succeed—both in getting us from fossil fuels to the renewables of the future *and* in preventing economic collapse in the process—businesses must be able to achieve fairly rapid returns on investment without massive capital costs. Moreover, most businesses need to be able to achieve such returns without waiting for the fruits of future technological progress that, contrary to traditional expectations, is not automatic. We have also argued that success in this unprecedented endeavor will require advances in the productivity not just of labor, but of the physical resources on which the whole global economy depends.

In the energy field, the most successful businesses of the coming years will not be those that pour money into the bottomless pits of "clean coal," thousand-mile pipelines, nuclear power plants, oil-drilling platforms, or corn ethanol production that consumes about as much fuel as it produces. The winners will be those that find ways to get more service out of each dollar of expenditure on existing sources of coal, oil, or natural gas, as well as on harnessing wind or solar power. We have sketched a series of strategies for doing so and have cited examples of companies that have made substantial gains with these strategies. But how do these gains translate to useful advice for business at large?

One way to answer that is to look a little more closely at the thinking of a few of the people who have pioneered these strategies so far. In the Introduction and Chapter 2, "Recapturing Lost Energy," we highlighted the case of Cokenergy, the rust-belt plant that a few years ago began converting waste heat from a coking facility into 90 megawatts (MW) of clean electricity per year to power the adjoining

Mittal Steel plant. That remarkable arrangement didn't come about simply because it was cutting-edge engineering; it also resulted from the willingness of far-seeing managers to question some previously unchallenged rules of the prevailing business culture and take a calculated risk. Cokenergy was the brainchild of an electric power engineer named Tom Casten, who recognized that if massive amounts of heat were being wasted all over the world, there was a large opportunity for providing a profitable energy service. Casten founded the company Primary Energy, Inc., and offered Mittal Steel an opportunity to save big on its fuel costs and carbon emissions. Primary Energy still runs that business today. We mentioned earlier that, in 2005, if the Cokenergy output of clean energy were added to that of a similarly innovative flare-gas recycling facility down the road at U.S. Steel, the two Indiana plants yielded more carbon-free electricity than the entire U.S. output of solar-photovoltaic electricity that year. By 2009, Primary Energy was generating 900MW of recycled energy from fossil-fuel waste in the steel industry alone.

In 2006, Casten became CEO of a new energy-recycling company, Recycled Energy Development (RED), which he now runs with his son, Sean. The industry he helped kick-start is growing fast. We return to his story shortly, but for now we want to focus on one key point. The Castens have kept in mind that they're not selling energy—they're selling energy service. Making their kind of service more profitable depends not on selling more fossil fuel, but on using less fuel for a given amount of useful work. It is an illustration of one of the core principles we describe in Chapter 10, "Policy Priorities." The incentives now prevailing throughout the global energy economy (to sell more coal, oil, cars, and power) need to be reversed. The Castens' services don't entail selling *any* fuel, because all the fuel their facilities use has already been purchased and burned. Incorporating the benefits of that approach into the management strategies of RED's client companies enables those companies to operate at a significantly higher level of productivity.

That leads us to three recommendations for business managers and investors who want to fare well in what is becoming an increasingly challenging environment.

1. Bring Energy Management to the Highest Level of Strategic Planning

The most successful managers of the coming years will recognize the importance of increasing *energy* productivity—not just *labor* productivity—in their internal operations. This will mean giving energy productivity a place at the top management table, along with human resources and financial management. It will mean recognizing energy service as a pillar of the core business, along with labor and capital. In essence, that is what the Indiana steel giants did.

In Chapter 4, "The Invisible-Energy Revolution," we cited a study by the Alliance to Save Energy which found that many executives have viewed energy-saving programs as just "technical" matters best left to engineers instead of constituting significant aspects of business strategy. Executives have typically assumed, "Energy is not our core business." In accordance with this now obsolescent assumption, energy has been given marginal attention and budgetary resources commensurate with its status as a secondary function; it's not seen as a *controllable cost of production and source of recoverable earnings.* We also recounted in Chapter 4 the story of a Dow Chemical Company experiment that dramatically challenged that "it's really not our business" assumption: the story of Ken Nelson and his midlevel engineers, who demonstrated for 12 consecutive years how an industrial plant could make significant productivity gains by focusing more intensively on energy management.

Unfortunately, in large companies, changing the hardened assumptions of management is much harder to accomplish than by just bringing another chair to the oval table. The difficulty is that industries and companies often have deeply established cultures and ideologies, and tend to strongly identify themselves with their core businesses. At Dow, where the core business is manufacturing plastics, it took more than 12 years for top management to recognize how the priorities were changing (perhaps partly because energy was fairly cheap during those years). For Cabot Corporation, the Louisiana company that considered building an energy-recycling facility but was foiled by a recalcitrant electric utility monopoly and public utility commission (see

Chapter 5, "The Future of Electric Power"), the core business was manufacturing and selling carbon black—and the company's will to see the project through eventually gave out. For General Motors, the core business was making and selling cars and trucks, not offering energy-efficient transportation; during the company's decades of decline, energy management was never a significant part of its strategy. It's difficult for a big company to change its core business.

The collapse of the U.S. auto industry in 2009 illustrates this point. At first, the public story was that the "Big Three" were desperately short of *capital,* and their CEOs were flying to Washington to ask for a handout. Then, as more of the story unfolded, it became clear that their *labor* situation was untenable as well: GM had a million people collecting benefits, but fewer than a tenth of that number were employed to cover those costs. The government attempted a temporary rescue, but the members of Congress—on both sides of the aisle—suspected that the first loan would have to be followed by another, and another, with no end in sight. No sign indicated that GM would be able to produce and sell more fuel-efficient vehicles in time to keep its ship afloat. As we now understand, the suspicions were warranted. An essential factor of economic growth in the U.S. auto industry—low costs of energy service, both in manufacturing motor vehicles and in the expected cost of driving those vehicles in the years to come—was missing.

So, granted that the idea of a core business is basic to a civilization built on specialization, it is still a fact that no business can be isolated from the legs that help it run. To suggest an analogy, a National Football League team owner might think of his superstar quarterback as the core of his team's public image and success on the field. Superstar quarterbacks are sometimes designated as "franchise" players and are accorded correspondingly elevated financial status. But when the game starts, the quarterback still has to be protected by a phalanx of powerful linemen. In a business, the three productivity factors of capital, labor, *and energy productivity* are that phalanx. If they're not quite the core output that interests stock market analysts, they're nonetheless indispensable to that output.

To take this analogy one step further, consider that, for the uncelebrated tackle or guard, as distinguished from the team's owner, the "core business" is to *enable* the quarterback to pass the ball. In other words, in the world of industry, the core can't work if the productivity legs of capital, labor, and energy aren't mobilized. In many industries, labor productivity has gotten all the attention, even as resource scarcities have grown and prices have risen, resulting in outsourcing and job exports. Energy productivity has been widely neglected, partly because energy costs have always been very low and partly because innovation has been foiled by the kinds of institutional barriers we have described in earlier chapters. For example, a firm hoping to profit from energy recycling can hire an energy-service company such as Primary Energy (for which *energy service* is the core business) to provide it with the missing third pillar of productivity. But the energy savings opportunity might still depend on making a deal with the local electric utility that has a legal monopoly on production and distribution. When that happens, it's as if the referee keeps ignoring late hits.

Elevating the energy-management leg to the same level of recognition and consideration in strategic planning as the labor and capital legs might neutralize the political bullying by utility lobbies and other guardians of the old regime, first for businesses whose managers can quickly grasp the importance of redefining how "core" business really works, and later for the entire energy economy. In short, real progress might be made toward doing business on a truly level playing field, with rules permitting real competition of the kind that can lower energy-service costs and increase productivity and profits, while at the same time reducing carbon emissions and earning greater public trust.

2. Recognize the Business Opportunities, and Risks, that Will Come with Rising Natural Resource Prices

One key to successful business planning in the coming years is to recognize that the perfect storm of post-peak-oil turbulence, obsolescent fossil-fuel technology, and escalating climate change is approaching, even as we begin to cross the energy-transition bridge.

A major effect will be rising natural resource prices, starting with those of oil and gas. This creates financial risks (and opportunities for innovative forms of risk spreading), to be sure, but it will also provide abundant opportunities for new technologies. We emphasized in the "Introduction" that the energy-transition strategy doesn't depend on new technology for immediate implementation, nor do individual companies need to depend on new technology to begin making short-term returns on energy investment. But that doesn't mean there should be any hesitation in the creative thinking, research, and development that could both strengthen the transition bridge and shorten the chasm it must cross.

At this point, it is important to be clear about the distinction between energy and energy service, because many businesses might soon see their respective prices begin to diverge, with make-or-break consequences. As we argued in Chapter 1 (and as our research on economic growth graphically suggests), future recovery and growth will most likely occur in businesses and sectors that manage to *reduce* the costs of energy service, even as the prices of fossil fuels erratically but inexorably *increase* all over the world.

In the past, under the boom conditions of the post–World War II era, it usually made sense for big manufacturing companies to wait for the inventors and "first movers" of innovative technologies to take the biggest risks, solve the fundamental technical and manufacturing problems, and gain a foothold in the marketplace before jumping in later with greater resources and economies of scale to reap the greater profits. This was the pattern in the semiconductor business, the computer business, and the biotech business.

But now an epochal change is in the wind worldwide, and the big companies—especially those in the United States—are unprepared to be first movers, even though late entrants might be too late. The near future will likely see a rapid move to plug-in hybrid cars, which will be succeeded in coming decades (in the cities) by electric vehicles (EVs), many of which will not be privately owned. Entrepreneurial and investment opportunities will abound in car-sharing software, EV infrastructure, e-cycles, battery manufacturing, bus rapid transit, and European- and Japanese-style high-speed rail. The housing and urban design sectors might see even greater opportunities for innovation, in zero-energy or low-energy architecture, construction, micropower, and urban preparation for climate change. Obvious—and some not-so-obvious—opportunities already exist in renewable energy, such as mapping and marketing sites for wind turbines or geothermal extraction.

All of these potential innovations will require capital, of course. The logical source of new capital for innovation is the "cash cows" of the economy: the oil and gas companies and the electric utilities. Those companies must eventually acknowledge that the high prices driving their hefty profits will also have an adverse effect on overall economic demand and employment. In principle, this fact should give them an incentive to keep prices down (by investing in alternatives), to keep demand up. In practice, the oil company executives are too easily seduced by the dream that there are still plenty of "gushers" to be found and that the near future will be like the past. Science and communication have important roles here. The U.S. government and the public will have to work hard to induce those ultraprofitable companies, starting with ExxonMobil, Shell, and BP to put some of their profits to work in ways that increase resource productivity and energy efficiency at the macroeconomic level.

We recognize that innovative entrepreneurs in the private sector have created much of the wealth in our society—a fact that has often been invoked as an argument against government investment in general: "Don't try to pick winners—let the market decide." Yet a careful review of history makes it clear that some infrastructure investments are too big or too slow to pay off in the private sector. The great hydroelectric projects (Boulder Dam, Grand Coulee Dam, the Tennessee Valley Authority), rural electrification, the Interstate highway program, nuclear power, jet engines, and computers are examples of technological progress initially funded by large government investments. The Internet began as a project of the Advanced Research Projects Administration (ARPA) in the Defense Department, intended to enable computers in major universities to exchange data. The government funds most university research. Fusion power and solar satellites or solar power from the moon are examples of projects with potentially enormous payback, but they're far beyond the capabilities of the private market.

One of the silver linings of the economic crisis could be that it is reminding the public that business, government, and civil society are not separate realms in a polarized struggle, but are highly—perhaps increasingly—interdependent. In earlier chapters, we described such public–private partnerships as the New York State Energy Research and Development Authority (NYSERDA), the European Passive House Project, and the World Alliance for Decentralized Energy (WADE). In our Endnotes and on our website, we list many more. The global crisis has increased the number of such alliances and the opportunities they create, for both technical assistance and financing. The World Business Council for Sustainable Development (WBCSD) and other progressive business groups have begun to emerge from the paralysis of anti-government polarization. In addition, growing alliances between private businesses and nonprofit organizations (NPOs) are offering guidance on what is environmentally sustainable, both for business and for society at large.

That brings us back to Tom and Sean Casten, who are entrepreneurs of the kind economists have in mind when they speak of "American ingenuity" or, on a more global scale, of "technological progress." They demonstrate the first two principles we recommend to business managers: They are acutely aware of the importance of both energy productivity and the abundance of business opportunity in that realm. Businesses like theirs might benefit from government boosts, but as public awareness of the new energy paradigm grows, they'll also get stronger support from private investment. In November 2007, the Castens' energy-recycling company announced that it had been approved to receive up to $1.5 billion in private investment from the Boston-based private-equity fund Denham Capital Management. Major investors included Harvard University and Bill Gates.

The Recycled Energy Development announcement was not an anomaly. As atmospheric levels of carbon dioxide continue to rise, along with public concerns about the global energy dilemma, private investment in the girders of the energy-transition bridge is shifting from tentative to robust. The key, as we have shown, is that such investments often can bring double dividends of corporate and social benefits, sometimes even at "negative cost." "[That discovery] is allowing much bigger capital deployments than we've seen," said

John Balbach, a managing partner at Cleantech Group LLC. of Brighton, Michigan at the time of the RED announcement. And as Riaz Siddiqui, a senior managing director at Denham, put it, "The exciting feature is reducing the carbon footprint of U.S. industry *profitably.*"

3. Get Ready, Wherever You Are

The coming energy transition will affect every kind of business, from coal mining to white-collar services (education, media, consulting, law, and so on), and from multinational corporations to one-person enterprises. Those who act most quickly to identify and adapt to the inputs will be most likely to survive and thrive. No business will remain unaffected, because no business can function without energy, whether in the form of food for its workforce, fuel for its factories, heating or lighting for its office space, or electricity for its telecommunications. In years past, most of these costs were relatively small and stable; they were not matters of great concern in planning business models or in satisfying lenders or investors. That reality is changing at every link in the economic chain. For example, a developer of office space can't just consider the cost of construction per square foot; To be competitive, he also needs to give considerable weight to the costs of all the building's energy services and to the security of the resources and technologies used to supply them. He must factor in his building's carbon footprint as well as its architectural footprint. That means climbing an architectural learning curve far higher than what sufficed for most of the past century. If the developer is particularly savvy, he will consider the building's proximity to public transit, car-sharing facilities, electric-vehicle infrastructure, bike lanes, and clean air.

Paleoanthropologists tell us that before civilization began, humans might have spent most of their waking hours hunting for or gathering food. The development of agriculture, with one farmer able to produce food for several other people, freed people to live in towns and develop other occupations and industries. But population expansion (more than a thousand-fold since the early civilizations) and voracious natural resource consumption have made the modern economy unsustainable, particularly in its heavy reliance on fossil fuels, which will likely continue precariously far into the present century. One consequence might be a reversion to the communal consciousness of a world in which all people share a common *pre*occupation. In the future, while we continue to further develop our specialized skills and businesses, our ability to affordably tap the physical energy on which the production of food and all other human activities and businesses depend might never be far from our consciousness.

12

How Much, How Fast?

This book makes a strong claim: that the key to the U.S. energy supply for the next generation—*until* renewables are up to scale—is *not* to develop new sources of oil or natural gas in ever harder-to-reach places at ever higher cost, but to greatly increase the amount of energy *service* we are getting from existing fossil-fuel sources. We further claim that this goal can be achieved with simultaneous sharp reduction of greenhouse gas emissions, relying on existing technologies and well-tested business strategies, and at surprisingly little cost to taxpayers. We see compelling reason to believe that such a strategy is crucial to sustaining economic health and civil order while providing a viable bridge to the clean-energy future.

We have reviewed the various "girders" needed to build the transitional bridge and have suggested that, together, they can increase the overall U.S. efficiency of converting primary energy to useful work from its present, anemic 13 percent to 20 percent or higher. That would be equivalent to decreasing the nation's energy consumption by more than half, without any reduction in our standard of living. By another measure, the American Council for an Energy-Efficient Economy (ACEEE) estimates that the United States can cost-effectively reduce the nation's energy consumption per dollar of GDP by 20–30 percent over the next 20–25 years. As we note in Chapter 5, "The Future of Electric Power," that estimate probably understates the possibilities. At this point, it's natural to wonder, just how *do* the energy savings (and reductions of carbon emissions) of our suggested girders add up? And how long will it take, both to build the bridge and to cross it?

To be candid, we have to begin our response to those questions by acknowledging that, to an unsettling degree, at this very turbulent moment in human history, those answers have not yet been determined. They depend on a range of events that no one can confidently predict. Whether the next decade will see one of the world's largest cities destroyed by a Category 5 hurricane; whether peak oil will prove to have been passed in 2010, 2012, or much later; whether some shocking geopolitical event on the scale of Pearl Harbor or 9/11 sweeps long-term planning from the consciousness of politicians and policymakers once again; whether the United States and other governments will have the political courage to remove long-standing institutional barriers to energy efficiency and establish effective incentives in their place; whether enough good fortune is encountered in the technological development of renewables (such as the discovery of a new source of tellurium for solar PV) to ramp up the new industries faster than anyone thought possible—these and other yet-to-be-answered questions will affect just how the benefits of the girders add up. The energy/climate future is a fast-moving target.

Additional uncertainties stem from the reality that the girders affect one another. If plug-in electric vehicles (EVs) make particularly rapid progress, for example, they will create new demand for electricity that could appear to justify new coal-burning power plants. On the other hand, an equally rapid penetration by local, decentralized CHP could compensate for the extra load on the grid from EVs. Moreover, a rapid buildup of plug-in EVs would provide storage capacity (on the grid itself) for intermittent sources such as wind or solar PV. Among the various girders, trade-offs will undoubtedly occur.

Nonetheless, it is possible to make fairly firm and meaningful estimates of the potential energy savings and emissions reductions for each of the eight girders, or for important components of them under specific scenarios. No person or government needs to see how they combine to grasp the value of acting on them individually, as critical economic and energy-security goals. It is enough to know that although uncertainties lie ahead, these girders support a common goal for the U.S. energy economy and for that of every country that now has a substantial fossil-fuel dependency. This chapter summarizes the benefits for the United States.

Waste-Energy Recycling in Industry

The first girder of the bridge is to "recycle" high-quality waste-energy streams from industrial plants, including petroleum refineries, natural gas compressors, coke ovens, carbon black plants, silicon refineries, glass furnaces, paper mills, and metallurgical operations of all kinds. As discussed in Chapter 2, "Recapturing Lost Energy," those streams can be in the form of high-temperature heat, steam, or flare-gas exhaust streams, or in the form of wasted compression energy. About 10,000MW (10GW) is currently recovered from these waste streams in the United States. But a study by the Lawrence Berkeley Laboratory for the Department of Energy has identified an additional 95,000MW (95GW) of potential that could generate as much as 10 percent of U.S. electricity at capital costs less than for new coal-burning power plants, with no fuel costs and with reduced carbon emissions. (A more recent EPA–DOE study set that potential lower, at 65GW, but still equal to about 7 percent of U.S. electric power output.) Most of those savings would be translated directly into reduced need for coal or natural gas. Another 6,500MW of carbon-free electricity could be generated by converting the compression energy in natural-gas pipelines to electricity by means of low-cost back-pressure turbines. Coal consumption, now almost entirely for electric power generation, could be cut by around 15 percent. Most of this could be done within ten years after legal and institutional constraints are removed.

Decentralized Combined Heat and Power (CHP)

The most wasteful use of energy in the United States today is as fuel for heating (and cooling) buildings—most scandalously, electric heating from fossil fuel–burning generating plants. As discussed in Chapters 5, 10, and 11, electric utilities have strong incentives to increase their capital investments in central plants and wires, and hence to sell more electricity, often offering discounts to builders to utilize resistive electric heat.

However, thanks to technological progress since the centralized power plant system became entrenched many decades ago, it is now practical to efficiently generate electricity on a relatively small scale (down to a few tens of kilowatts), suitable for apartment houses, hospitals, schools, department stores, small factories, and even individual houses. To convey a general sense of the magnitude of savings that could be achieved by shifting all new capacity to CHP, we note that in 2008, the World Alliance for Decentralized Energy (WADE), using the International Energy Agency's base data, estimated that capital savings, worldwide, could amount to $5 trillion over the next 20 years. If the United States accounts for anything close to its current one-fifth of global energy consumption (though that share is likely to fall, as China and other countries grow faster), we can expect that the potential savings for U.S. utility investments would amount to more than $1 trillion during that period. Reductions of fossil-fuel use would be commensurate.

To offer another broad perspective, note that the United States currently gets only 8 percent of its electric power from CHP plants, but several countries exploit CHP far more intensively: Denmark, 51 percent; Finland, 37 percent; and Russia, 31 percent. A significant part of the heat from CHP in those countries does go to district heating, which is of little use in the United States. But that hardly diminishes the fact that modern CHP could—and does, in those countries—provide far greater energy service per unit of fuel than the U.S. industry does. Again, the benefits for carbon emissions are commensurate: CO_2 emissions from Denmark declined from 60 million metric tons in 1991 to about 50 million tons in 2005, a period during which emissions in most other countries rose.

In a paper published in *American Scientist* in 2009, Thomas Casten and Philip Schewe estimated the *immediately* profitable savings potential for decentralized CHP in the United States to be 135GW, which, added to the 65GW (or greater) potential from industrial energy recycling, amounts to 20 percent or more of total U.S. power output and a similar percentage reduction in carbon-dioxide emissions. These numbers also don't take into account the potential contributions from micropower (rooftop PV, small wind-turbines, and so on), discussed shortly, nor do they account for the probability that fossil energy prices will inexorably rise in the future, making efficiency more profitable. We believe that the average exergy (useful energy) efficiency of electricity power generation in the United States could increase from 33 percent to 40 percent in the next 10 years and to 50 percent by midcentury, simply by encouraging decentralized CHP.

Energy Use Efficiency in Industry and Buildings

Industry consumes about a quarter of all energy consumed in the United States. Buildings consume another quarter, not including electricity consumed inside buildings. As discussed in Chapter 5, "The Future of Electric Power," we strongly disagree with the view of mainstream economists that the potential for cutting energy use and emissions in this sector is relatively insignificant. The American Council for an Energy Efficient Economy (ACEEE) issued a study in 2008 finding that three-fourths of all new U.S. energy demand over the previous 38 years had been met by increasing efficiency, and only one-fourth by new supplies of oil, coal, or other conventional sources. About 72 percent of that "invisible energy" boom took place in the buildings and industry sectors.

According to the Environmental Protection Agency (EPA) and the Vanguard Group of mutual funds, efficiency investments are generally low in risk (almost as good as government T-bills) yet have produced high returns averaging around 25 percent, far higher than most stocks or bonds. That assessment predated the financial meltdown of 2008–2009. However, given the role of energy services in economic growth, as discussed in Chapter 1, "An American Awakening," the returns on efficiency investments, relative to other investment options, could be even higher in coming years.

The ACEEE notes that "we have barely begun to scratch the surface of the potential savings that additional investments in energy efficiency technologies could provide" and that "our research findings indicate that in an environment of accelerated market transformation and rapid growth in efficiency investments, total investments in more energy efficiency technologies could increase the annual energy efficiency market by nearly $400 billion by 2030, resulting in an annual efficiency market of more than $700 billion in 2030."

End-Use Efficiency

We have given end-use efficiency relatively little attention in this book, except in our discussion of cars in urban environments, because it is the one of the strategies most widely discussed in popular media. We infer that most Americans are now aware of the benefits of energy-efficient lighting, refrigerators, air-conditioners, and fuel-efficient cars. At this point, however, we should acknowledge that this girder is by no means a small part of the picture. In its 2008 analysis of the U.S. energy efficiency market, the ACEEE found that appliances and electronics benefited from the largest share of efficiency-related investments over the previous several decades: 29 percent, or $87 billion. (Interestingly, appliances got more attention than cars.)

In another 2008 study, the International Energy Agency (IEA) concluded, "Improved efficiency and decarbonizing the power sector could bring emissions back to current levels by 2050." Over that four-decade span, the IEA estimated that end-use fuel efficiency would account for 24 percent of the projected emissions reductions, and end-use electricity efficiency would account for 12 percent. Significantly, even after four decades, the combined end-use gains (36 percent of the total) would still be substantially larger than those provided by renewables (21 percent). This is another confirmation of how critical it is to have a reliable transition bridge to the day when renewables achieve real primacy. However, we think that the use of renewables can and should grow faster than the conservative IEA projections.

Sometimes simple solutions trump complex ones, and that might be true for automotive efficiency during the energy-transition period. One of the authors currently drives a modestly priced five-seat European diesel car that gets 40 miles per gallon, mainly for urban trips. If everyone in the United States got that level of fuel economy, U.S. motor fuel consumption would decrease by nearly half.

Despite all the attention efficiency has received, it is hard to make realistic projections of future efficiency gains, in view of the economic shock to the global economy and in view of our own cautions regarding the forecasting of economic growth (discussed in Chapter 1). On the other hand, the shift of energy and climate priorities that occurred with the election of President Obama initiated a program of long-overdue energy subsidies emphasizing efficiency and conservation, as well as renewables. In 2007, federal subsidies for end use amounted to just $2.2 billion. With the "incentive" package of 2009, we started getting more serious. In the coming years, if substantially larger subsidies are directed to truly productive strategies such as integrating efficient appliances with zero-energy, zero-emissions houses, or resuming progress in CAFE standards for cars and trucks, the payback could be substantial and relatively rapid. Given the critical relationship between the (low) costs of energy services and the prospects for economic productivity and growth, these more robust investments could play a key role in repowering the U.S. economy in the next decade.

Kick-Starting the Micropower Revolution

Assuming the recommended changes in PURPA and state laws regulating electric utilities, we will see a lot of retrofitting of medium-scale CHP systems for shopping centers, schools, hospitals, office buildings, apartment buildings, and small factories. We'll probably also see a few tentative applications of very small CHP systems to new houses (although the "passive house" concept discussed in Chapter 8, "Preparing Cities for the Perfect Storm," makes home-heating applications in new construction unnecessary). However, rooftop PV can potentially convert rooftops and other flat south-facing surfaces all over the country into a major source of clean electricity (as well as hot water) for the grid long before the end of this century. More to the point, this development is already well underway in a number of countries and states (admittedly, with the help of subsidies, but without the need to break the utility monopoly), and it has created excitement and a dynamic new industry. California's Million Solar Roofs Plan is expected to produce 3,000MW of electricity and reduce greenhouse gases by 3 million tons by 2018, equivalent to taking a million cars off the road.

Only a few other states have the sunshine Southern California has, but even if the other 49 states could tap solar PV at an average of one-quarter the California rate per capita, the national gain would be roughly twice the California gain. More significant savings would take another decade or two. However, even in the nearer term, the contribution of rooftop PV can help provide power for a network of charging stations for electric vehicles, thus reducing or eliminating the need for new central power plants for that market.

Substituting Energy Services

As noted in our policy discussion in Chapter 10, a core principle of the bridge strategy is to restructure energy management so that, instead of having incentives to sell more products (and, therefore, more of the energy used to produce or operate those products), businesses would sell energy services that become more profitable if they use *less* energy. We noted that a principal mechanism for such substitution might be Energy Service Companies (ESCs), which provide such services as home heating or lighting instead of providing electric power. If upgrading residential energy efficiency can reduce fossil-fuel consumption by an average of 20 percent, that margin would be enough to provide incentives to both ESCs and consumers to make those upgrades. But even apart from ESCs, intelligent substitutions can have major payoffs. For example, if Internet shopping and telecommuting together could replace 10 percent of all automotive trips within the next ten years, they would reduce U.S. oil consumption by about two million barrels a day. That alone would reduce the need for imported oil by one-seventh.

Redesigning Cities for the Future

This is one area in which the kinds of uncertainties described at the beginning of this chapter make quantification particularly difficult. We can compare the benefits of investing in urban design for a future of disrupted climate and resource scarcities to the benefits of buying insurance: We can't predict the outcome for any particular city or span of time, but actuarial data say it's an investment we'd be foolish not to make.

To put the assessment in perspective, consider what happened to Florida in 1992, when Hurricane Andrew struck. Damages amounted to some $38 billion, and some insurance companies went out of business. If that hurricane had hit Miami head-on, the damage would have been comparable to that of an atomic bomb. Thirteen years later, Hurricane Katrina inflicted damages of $40 billion. With the onset of climate change, the risks of such events grow larger with every passing year. If a national (or global) program of climate-change mitigation and adaptation is implemented, averting the worst of even one major hurricane or storm surge will have made the investment worthwhile. But the likely reality for the next century is that we will see *many* extreme-weather events, of varying degrees of devastation. Comprehensive programs of combined climate mitigation and adaptation have become matters of national security as well as energy security.

It's not just coastal or riverside cities that can—and should—be gradually redesigned for climate change, however. Programs aimed at reducing automotive domination of urban spaces, increasing reliance on bus rapid transit, and reducing or eliminating the need for energy to heat buildings (as discussed in Chapter 8) will generate long-term savings that, in time, could more than cover the costs of the upgrading.

Reforming Fresh-Water Management

Fresh-water conservation is important in its own right, and doesn't need to be defended in terms of energy or climate implications. Nearly all economists agree on the need for rationalizing water markets. The usual prescription is putting a price on fresh water. The problem is that most fresh water is used for irrigation, and large-scale food production depends on large-scale irrigation. But industrial and urban water users are generally prepared to outbid farmers for irrigation water. Expanding cities in dry areas such as Southern California, Nevada, Arizona, and Texas have diverted huge amounts of water from the few rivers in the region (especially the Colorado), leaving many farmers relying on wells with inefficient pumps and sprayers. Much of this water comes from aquifers that will run dry in a few decades, at most, resulting in a possible replay of the great Dust Bowl of the early 1930s (except that, the next time, there will be nowhere for the next generations of "Okies" to go—California, the main 1930s destination, is now in water deficit itself). Because water scarcities are exacerbated by the need for pumping water from ever-increasing depths and distances, they impose greater demands on energy. The current practice of using large amounts of irrigation water to grow corn or soybeans for the ethanol used to replace small amounts of gasoline further worsens the conundrum. Each gallon of eliminated ethanol production would save 10,000 gallons of water—and the energy used to pump it.

Investment in the Future: The Make-or-Break Moment

Even under what President Obama called in his inaugural speech the "darkening clouds" of the global future, the way to that future is now clear: Our investments in that future must be *twofold,* with one stream directed to the clean-energy, low-carbon economy that will phase in over the next several decades, and another, equally strong, directed to the transitional bridge. Without adequate attention to the bridge, the American and global economies could collapse under the mounting pressures of rising population, resource scarcities, environmental decline, and climatic disruption.

But we now have abundant indications that a safe and strong bridge *can* be built, both quickly and affordably, if the kinds of policies outlined here are adopted. Among the signs is a U.S. power-generation system that for the past four decades has been stuck at 33 percent efficiency, when existing technologies unleashed by regulatory changes can raise that to 60 percent or higher in time. We have an opportunity to save hundreds of billions of dollars in capital costs of new electric power plants by decentralizing an obsolescent central plant system. We have an opportunity to reduce U.S. fossil fuel use and carbon emissions 10 percent or more by harnessing waste-energy streams. We have a momentous opportunity to raise the *overall* energy efficiency of the U.S. economy, now stalled at 13 percent, to at least 20 percent efficiency, and probably higher. And, of course, we have an opportunity to save the cities and towns of the East Coast and Gulf Coast from the potentially catastrophic impacts of intensifying climate change—and, at the same time, to materially upgrade the energy security and quality of life in all other cities and towns.

Perhaps most significant from a pragmatic standpoint, with existing, well-tested, and relatively inexpensive methods, we have the immediate opportunity—not theoretical, not "someday," not "if only"—to do all this in ways that provide abundant opportunities for profitable investment. Finally, this is an opportunity—an imperative, we believe—to take a course of action that, far from harming or further depressing the economy, as mainstream economists have feared, will give it robust new life.

Comments and References

Introduction: The Chasm to Be Crossed

Can renewables replace coal in a decade? Al Gore's petition to repower America (see http://wecansolveit.org) identifies wind power, solar thermal power, solar photovoltaics (PV), and geothermal power as options capable of replacing all coal-fired power plants within a decade (Gore, 2008) That goal is unrealistic. Granted, U.S. production of renewables grew fast in the years before the financial crash of 2008, from 352 terawatt-hours (TWh) in 2006 to about 500TWh in 2008, an eighth of the U.S. total. But 80 percent of that renewable energy produced now is hydroelectricity, which can't grow (virtually all the possible sites are already in use); much of the rest comes from burning wood and municipal waste. The sources we (and Gore) count on for the future provide less than 2 percent of U.S. electricity production now (United States Department of Energy annual). For further discussion of the prospects for **wind, solar-photovoltaic, solar thermal,** and **geothermal** energy, see the book's website at www.informit.com/register. (Go to this URL, sign in, and enter the ISBN. After you register, a link to the source content will be listed on your Account page, under Registered Products.)

Rust Belt factories in Indiana: See (Casten and Ayres, 2007).

Energy, exergy, efficiency, and useful work: Energy is one of those slippery concepts that everybody uses without a thought, but hardly anybody really understands. It has been called "the ultimate resource," which is not a definition. However, for purposes of this book, it is important to realize that most people use the term *energy* when they really mean *exergy*. *Exergy* is the technical term for that component of energy that is potentially available to do *useful work* (Rant, 1956; Glansdorff, 1957). Several kinds of exergy exist: mechanical work, electrical work, chemical work, and so on. The technical definition of *work* gets us back to thermodynamics, but the basic idea is that mechanical work is needed to accelerate a vehicle or to over-come friction, air resistance, or gravity (for example, to climb a hill or lift a bucket); electrical work is needed to overcome electrical resist-ance; chemical work is needed to reduce an ore or separate the ele-ments in a compound, and so on. For a comprehensive discussion, see (Szargut, Morris, and Steward, 1988).

- Casten, Thomas R., and Robert U. Ayres. "Energy Myth #8: The U.S. Energy System Is Environmentally and Economi-cally Optimal." In *Energy and American Society: Thirteen Myths*, edited by B. Sovacool and M. Brown. New York: Springer, 2007.
- Glansdorff, P. "On the Function Called Exergy and Its Applica-tion to Air Conditioning." *Bull. Inst. Int. Froid* Supp 2: 61–62.
- Gore, Al. *A Generational Challenge to Repower America.* Cited 31 July 2008. Available from www.wecansolveit.org/pages/.
- Rant, Z. "Exergy, a New Word for Technical Available Work." *Forsch. Ing. Wis.* 22 (1): 36–37.
- Szargut, Jan, David R. Morris II, and Frank R. Steward. *Exergy Analysis of Thermal, Chemical, and Metallurgical Processes.* New York: Hemisphere Publishing Corporation, 1988.
- United States Department of Energy, Energy Information Administration. *EIA Annual Energy Review.* Washington, DC: United States Government Printing Office.

1. An American Awakening

Historical statistics. For historical data since 1970, we use (United States Department of Energy, annual). For earlier years, we use (United States Bureau of the Census, 1975) and (Schurr and Netschert, 1960). For data on **energy prices,** apart from Schurr, et al., and the annual publications of the Energy Information Agency in the U.S. Department of Energy (United States Department of Energy, annual), see also (Potter and Christy, 1968).

Energy and economic growth: The first to call attention to the fundamental link between energy availability and economic growth was British Nobel Prize–winning chemist Frederick Soddy, in the 1930s. Most economists dismissed Soddy as a crank, although most of his recommendations have since been adopted. His sin, in the eyes of economists, was to advocate an energy theory of value and to suggest that the money supply be tied to energy availability (Soddy, 1933, 1935). The first economists to emphasize the fundamental importance of the laws of thermodynamics in the economic system were Nicholas Georgescu-Roegen and Herman Daly (Georgescu-Roegen, 1971, 1979; Daly, 1979). They saw the economic system as a sort of living organism extracting high-quality (low-entropy) resources from the environment and excreting low-quality (high-entropy) wastes that cannot be recycled indefinitely without a continuous supply of solar or fossil energy. This view is inconsistent with the standard theory of economic growth, promulgated back in the 1950s by Nobel Prize–winner Robert Solow (for example, in Solow, 1957, 1956), in which most growth is attributed to unexplained "technological progress" (or "total factor productivity") but energy consumption plays no explicit role. After the "energy crisis" of 1973–1974, several economists in the 1970s tried to explain economic growth quantitatively in terms of a production-function approach, including notably primary energy inputs (Jorgenson. 1978, 1984). But this approach met with limited success, thanks to the widely assumed "cost-share" constraint (that the marginal productivity of energy must be equal to its cost share in the national accounts), which implied that energy could not be very important because of its minuscule cost share (Denison, 1979, 1985). The first to succeed in explaining past growth of the U.S., U.K., German, and Japanese economies, for brief periods since 1970 (by ignoring this supposed cost-share constraint, which was derived from an ultrasimplistic single-sector, single-product economic model) was German physicist Reiner Kuemmel and his colleagues (Kuemmel 1982, 1989; Kuemmel and Lindenberger, 1998). Kuemmel has subsequently shown (see www.informit.com/register) that the cost share constraint is not applicable in a realistic multisector, multiproduct model. The last step was to explain "technological progress" in terms of the increasing technological efficiency of converting primary energy into useful work (mechanical work, chemical work, electrical work, and so on) (Ayres, Ayres, and Warr, 2003). Inserting "useful work" in place of primary energy in Kuemmel's production function has successfully explained economic growth since 1900 for the United States, and later for Japan, the United Kingdom, and Austria (Ayres and Warr, 2005, 2009). For further discussion, see www.informit.com/register.

Economic growth theory—the need for a new approach: See (Ayres, 1998).

Effects of oil shocks: See (Olson, 1988; Zivot and Andrews, 1992; Hamilton, 2005, 2003; and Roubini and Setzer, 2004).

The cost share theorem and the need for a three-factor model: See (Kuemmel, et al., 2008) and (Kuemmel, Ayres, and Lindenberger, 2008) and www.informit.com/register.

Peak oil: This phenomenon is known in some quarters as the Hubbert peak (Hubbert, 1956; Hubbert, 1962; Hubbert, 1969, 1973). It has both adherents and skeptics. The skeptics are mostly mainstream economists who believe, as a matter of faith, that modestly higher prices will automatically call forth new supply sufficient to meet growing demand. Nevertheless, the empirical evidence seems increasingly to favor the peak oil adherents, as reflected in sharp changes in the IEA outlook since 2004. The sharply rising demand from China and India in 2006–2007, together with the increasing gap between consumption, depletion, and discovery, is almost certainly the primary reason for the rather sudden surge in prices in 2008. For further reading, see (Hatfield, 1997, 1997; Campbell, 2004; Deffeyes, 2001; Strahan, 2007; and Deffeyes, 2005).

Scientists' warning: See (Kendall, 1992). The signatories included more than 100 Nobel Prize winners, including 97 in physics, chemistry, and medicine, and 7 in economics

Red List of threatened species: See (Baillie, Hilton-Taylor, and Stuart, 2004; and Baillie and Groombridge, 1996).

Fastest mass extinction: See (American Museum of Natural History; Futter, 1998).

Energy myths: See (Sovacool and Brown, 2008).

- Ayres, Robert U. "Towards a Disequilibrium Theory of Economic Growth." *Environmental and Resource Economics* 11, special issue 3/4 (1998): 289–300.

- Ayres, Robert U. *The Economic Growth Engine: How Energy and Work Drive Material Prosperity.* Cheltenham, U.K., and Northhampton, Massachusetts: Edward Elgar Publishing, 2009.
- Ayres, Robert U., Leslie W. Ayres, and Benjamin Warr. "Exergy, Power, and Work in the U.S. Economy, 1900–1998." *Energy* 28, no. 3 (2003): 219–273.
- Ayres, Robert U., and Benjamin Warr. "Accounting for Growth: The Role of Physical Work." *Structural Change & Economic Dynamics* 16, no. 2 (2005): 181–209.
- Baillie, J. E. M., and B. Groombridge, eds. *1996 IUCN Red List of Threatened Animals.* Gland, Switzerland: International Union for the Conservation of Nature, 1996.
- Baillie, J. E. M., C. Hilton-Taylor, and S. N. Stuart, eds. *2004 IUCN Red List of Threatened Animals.* Gland, Switzerland: International Union for the Conservation of Nature (IUCN), 2004.
- Campbell, Colin J. *The Coming Oil Crisis.* Brentwood, U.K.: Multi-Science Publishing Co., 2004
- Daly, Herman E. "Entropy, Growth, and the Political Economy." In *Scarcity and Growth Reconsidered,* edited by V. K. Smith. Baltimore: Johns Hopkins University Press, 1979.
- Deffeyes, Kenneth S. *Beyond Oil: The View from Hubbert's Peak.* Hardcover ed. Princeton, New Jersey: Princeton University Press, 2001.
- Deffeyes, Kenneth S. *Beyond Oil: The View from Hubbert's Peak.* Hardcover ed. Hill and Wang, 2005.
- Denison, Edward F. "Explanations of Declining Productivity Growth." *Survey of Current Business* 59, Part II (1979): 1–24.
- Denison, Edward F. *Trends in American Economic Growth, 1929–1982.* Washington, DC: Brookings Institution Press, 1985.
- Futter, Ellen V., et al. "Biodiversity in the Next Millennium." New York: American Museum of Natural History and Louis Harris and Associates Inc., 1998.
- Georgescu-Roegen, Nicholas. *The Entropy Law and the Economic Process.* Cambridge, Massachusetts: Harvard University Press, 1971.

- Georgescu-Roegen, Nichols. "Energy Analysis and Economic Valuation." *Southern Economic Journal* (April 1979): 1023–1058.
- Hamilton, James D. "What Is an Oil Shock?" *Journal of Econometrics* 113 (2003): 363–398.
- Hamilton, James D. "Oil and the Macroeconomy." In *The New Palgrave: A Dictionary of Economics*, edited by J. Eatwell, M. Millgate, and P. Newman. London: Macmillan, 2005.
- Hatfield, Craig Bond. "How Long Can Oil Supply Grow?" Golden, Colorado: M. King Hubbert Center for Petroleum Supply Studies, Colorado School of Mines, 1997.
- Hatfield, Craig Bond. "Oil Back on the Global Agenda." *Nature* 387 (May 1997): 121.
- Hubbert, M. King. "Nuclear Energy and the Fossil Fuels." Houston, Texas: Shell Development Corporation, 1956.
- Hubbert, M. King. "Survey of World Energy Resources." *The Canadian Mining and Metallurgical Bulletin* 66, no. 735 (1973): 37–54.
- Hubbert, M. King. "Energy Resources: A Report to the Committee on Natural Resources of the National Academy of Sciences—National Research Council." Washington, DC: National Research Council/National Academy of Sciences, 1962.
- Hubbert, M. King. "Energy Resources." In *Resources and Man*, edited by Cloud. San Francisco: W. H. Freeman and Company, 1969.
- Jorgenson, Dale W. "The Role of Energy in the U.S. Economy." *National Tax Journal* 31 (1978): 209–220.
- Jorgenson, Dale W. and Barbara M. Fraumeni. "The Role of Energy in Productivity Growth." *The Energy Journal* 5, no. 3 (1984): 11–26.
- Kendall, Henry. "World Scientists' Warning to Humanity." *Union of Concerned Scientists*, 18 November 1992.
- Kuemmel, Reiner. "The Impact of Energy on Industrial Growth." *Energy* 7, no. 2 (1982): 189–201.
- Kuemmel, Reiner. "Energy As a Factor of Production and Entropy As a Pollution Indicator in Macroeconomic Modeling." *Ecological Economics* 1 (1989):161–180.

- Kuemmel, Reiner, and Dietmar Lindenberger. "Energy, Technical Progress, and Industrial Growth." Paper read at Advances in Energy Studies: Energy Flows in Ecology and Economy, Porto Venere, Italy, May 1998.

- Olson, Mancur. "The Productivity Slowdown, the Oil Shocks, and the Real Cycle." *Journal of Economic Perspectives* 2, no. 4 (1988): 43–69.

- Potter, Neal, and Francis T. Christy, Jr. *Trends in Natural Resource Commodities.* Baltimore: Johns Hopkins University Press, 1968.

- Roubini, Nouriel, and Brad Setzer. *The Effects of the Recent Oil Price Shock on the U.S. and Global Economy.* New York University, 2004. Available from www.stern.nyu.edu/globalmacro/ OilShockRoubiniSetzer.pdf.

- Schurr, Sam H., and Bruce C. Netschert. *Energy in the American Economy, 1850–1975.* Baltimore: Johns Hopkins University Press, 1960.

- Soddy, Frederick. "Wealth, Virtual Wealth, and Debt." In *Masterworks of Economics: Digests of 10 Classics.* New York: Dutton, 1933.

- Soddy, Frederick. *The Role of Money.* New York: Harcourt, 1935.

- Solow, Robert M. "A Contribution to the Theory of Economic Growth." *Quarterly Journal of Economics* 70 (1956): 65–94.

- Solow, Robert M. "Technical Change and the Aggregate Production Function." *Review of Economics and Statistics* 39 (August 1957): 312–320.

- Sovacool, Benjamin K., and Marilyn A. Brown, eds. *Energy and American Society: Thirteen Myths.* New York: Springer, 2008.

- Strahan, David. *The Last Oil Shock.* London: John Murray Ltd., 2007.

- United States Bureau of the Census. *Historical Statistics of the United States, Colonial Times to 1970.* Bicentennial ed. 2 vols. Washington, DC: United States Government Printing Office, 1975.

- United States Department of Energy, Energy Information Administration. *EIA Annual Energy Review.* Washington DC: United States Government Printing Office.

- Zivot, E., and Donald. W. K. Andrews. "Further Evidence on the Great Crash, the Oil Shock, and the Unit Root Hypothesis." *Journal of Business and Economic Statistics* 10, no. 3 (1992): 251–270.

2. Recapturing Lost Energy

Cokenergy and Mittal Steel: See (Casten and Ayres, 2007).

Kodak plant: See (Casten and Ayres, 2007).

U.S. Solar PV Production: See (Dorn, 2007; Worldwatch, 2008).

Combined Heat and Power (CHP): This strategy has been in limited use for several decades (Newman, 1997; U.S. DOE, 1999). But CHP was not widely recognized as having significant potential for the energy economy until February 2006, when the International Energy Agency (IEA) in Paris held its first international meeting to initiate a research program—but with no research budget except what interested companies could provide. However, in July 2007, the program was incorporated into the IEA's G8 Programme of Work on Climate Change and Clean Energy. A series of reports on work in progress were prepared and published in 2008 (International Energy Agency, 2008; Tanaka, 2008). Also see (Casten and Schewe, 2009).

- Casten, Thomas R., and Robert U. Ayres. "Energy Myth #8: The U.S. Energy System Is Environmentally and Economically Optimal." In *Energy and American Society: Thirteen Myths*, edited by B. Sovacool and M. Brown. New York: Springer, 2007.

- Casten, Thomas R., and Philip F. Schewe. "Getting the Most from Energy." *American Scientist* 97 (January/February, 2009): 26–33.

- Dorn, Jonathan G. *Solar Cell Production Jumps 50 percent in 2007.* Earth Policy Institute, 27 December 2007. Cited 14 January 2009. Available from www.earth-policy.org/Indicators/Solar/2007.htm.

- International Energy Agency. "Combined Heat and Power: Evaluating the Benefits of Greater Global Investment," edited by T. Kerr. Paris: International Energy Agency (IEA), 2008.

- Newman, John. "Combined Heat and Power Production in IEA Member Countries." In *Cogeneration: Policies, Potential, and Technologies*, edited by P. K. Dadhich. New Delhi, India: Tata Energy Research Institute (TERI), 1997.
- Tanaka, Nobuo. "Today's Energy Challenges: The Role of CHP." Paris: International Energy Agency (IEA), 2008.
- U.S. DOE. "Review of Combined Heat and Power Technologies." ONSITE SYCOM Energy Corporation for the California Energy Commission with the U.S. Department of Energy, Office of Energy Efficiency and Renewable Energy, 1999.
- Worldwatch Institute. "U.S. Solar PV Production." In *Vital Signs 2007–2008*, edited by Worldwatch Institute. San Francisco: W. W. Norton & Co, 2008.

3. Engineering an Economic Bridge

Updated forecasts of climate change: See (Clark and Weaver, 2008).

Incomes of the top 1 percent vs. bottom 90 percent: See (Huang and Gum, 1991). The authors analyzed IRS data from 1913 through 2006 and found that income concentration in 2006 was at its highest level since 1928.

Externalities and GDP: Externalities are a general economic term for losses (or gains) to "third parties"—that is, to bystanders who are not directly involved in an economic transaction with others. Economic models based on exchange transactions for goods and services with prices have great difficulty dealing with externalities because no market mechanism determines prices or damages. See (Ayres and Kneese, 1969). See also "Measures of welfare vs. growth" under the notes for Chapter 1.

Eight girders of the energy-transition bridge:

(1) **Recycling waste energy streams** and (2) **combined heat and power:** See notes for Chapter 2; (Casten and Ayres, 2007); and personal communication from Tom Casten, Chairman, Recycled Energy Development, January 2009.

(3) **Increasing energy efficiency in industrial processes and buildings** and (4) **increasing energy efficiency in consumer end uses:** For numerous examples, see (Lovins, et al., 1981; Lovins, 1986, 1996; and von Weizsaecker, Lovins, and Lovins, 1998). See also notes for Chapters 4 and 8.

(5) **Decentralizing electric power:** See notes for Chapter 6.

(6) **Substituting energy services** and (7) **redesigning buildings and cities for climate change:** See notes for Chapters 7 and 8.

(8) **Reforming fresh-water management strategies:** See notes for Chapter 9 and (Wilkinson, 2008).

- Ayres, Robert U., and Allen V. Kneese. "Production, Consumption, and Externalities." *American Economic Review* 59 (June 1969): 282–297.

- Clark, P. U., and A. J. Weaver, et. al. "Abrupt Climate Change." In *Report by the U.S. Climate Change Science Program and the Subcommittee on Global Change Research.* Reston, Virginia: United States Geological Survey, 2008.

- Huang, Dennis B. K., and Burel Gum. "The Causal Relationship Between Energy and GNP: The Case of Taiwan." *Journal of Energy & Development* 16, no. 2 (1991): 219–226.

- Lovins, Amory B. *State of the Art in Water Heating.* Snowmass, Colorado: Rocky Mountain Institute, 1986.

- Lovins, Amory B. "Negawatts: Twelve Transitions, Eight Improvements, and One Distraction." *Energy Policy* 24, no. 4 (1996): 331–343.

- Lovins, Amory B., L. Hunter Lovins, Florentin Krause, and Wilfred Bach. *Least-Cost Energy: Solving the CO$_2$ Problem.* Andover, MA: Brickhouse Publication Co., 1981.

- U.S. House of Representatives. Committee on Science and Technology, Subcommittee on Energy and the Environment. *Testimony of Robert Wilkinson, Ph.D.* 14 May 2008.

- von Weizsaecker, Ernst Ulrich, Amory B. Lovins, and L. Hunter Lovins. *Factor Four: Doubling Wealth, Halving Resource Use.* London: Earthscan Publications Ltd., 1998.

4. The Invisible-Energy Revolution

The "invisible-energy" boom: See (Ehrhardt and Laitner, 2008).

The Bridges Report and its successors: See (Bridges, 1973). In the early 1970s, the exergy ("second law") in terms of the efficiency of water heaters and fuel-burning space-heating systems generally was (and still is) only around 10 percent or less; the efficiency of electric heating is even lower, due to the losses in the generating plant. In exergy terms, much the same problem applies to all of Bridges's calculations. Using the numbers from the APS study (Carnahan, et al., 1975), together with energy use data from standard Census sources, the real exergy efficiency of the United States in 1973 was far lower than Bridges's figure, especially if one takes into account exergy losses by electrical devices in end uses (Ayres, Narkus-Kramer, and Watson, 1976). These calculations have been refined and updated since then (Ayres, 1989; Nakicenovic, Gilli, and Kurz, 1996). When these calculations are applied to the whole economy, it turns out that the exergy efficiency of the U.S. economy in 1973 was about 10 percent. Today, using the same methodology, the efficiency of the U.S. economy is about 13 percent (Ayres, Ayres, and Warr, 2003). These revisions have profound implications for the future: They imply that, even though gasoline engines and steam turbines are near their ultimate limits, plenty of room still exists to improve the exergy efficiency of the U.S. economy as a whole.

Calculation of (exergy) efficiency: The exergy efficiency of a process is the ratio of useful work output to exergy input. It is sometimes called "second-law efficiency," to distinguish from "first-law efficiency," which is a misleading measure. An example of the latter is the fraction of the heat from a combustion process that is transferred to a heat exchanger (such as a radiator), as opposed to the fraction lost to the environment through the smokestack. Detailed calculations of the second-law efficiency of many familiar conversion operations—from internal combustion engines to refrigerators and air-conditioners—were carried out, compared, and presented in a common format by a 1975 summer study sponsored by the American Physical Society (Carnahan, et al., 1975).

S.P. Newsprint: Jan Schaeffer and Scott Conant, "SP Newsprint Reaps Multiple Benefits from Energy Upgrade" (press release), SP Newsprint Company and Energy Trust of Oregon, Inc.; 7 June 2006.

University of Cincinnati energy savings: Facility Management Department, University of Cincinnati, "Catalyzing the Natural Linkage of Energy, Economics, and Environment" (press release), Office of Energy Efficiency, Community Development Division, Ohio Dept. of Development; 2008.

J. R. Simplot potato-processing plant savings: See (Hawk, 2006).

Trillion calculations per second: Intel Corp (press release) and Scott Jagow, "Marketplace" (interview), American Public Media, 12 February 2007. BBC News noted at the time of the Intel announcement that Intel had achieved a teraflop performance the first time; 11 years earlier, at Sandia National Laboratory, it had required "a machine that took up more than 2,000 square feet, was powered by nearly 10,000 Pentium Pro processors, and consumed more than 500 kilowatts of electricity."

New York State Energy R&D: See (Ferranti, et al., 2000).

Pollution prevention and energy efficiency: Chloe Birnel, "What's New in P2" (Pacific Northwest Pollution Prevention Resource Center, 1999, 2000, 2009). www.pprc.org/news.

Dow Chemical Company's energy efficiency contest: See (Nelson, 1993). Consider this important coda to this story: In November 2008, one week after the election of Barack Obama (who had made it clear that he did not share the Bush–Cheney administration's dismissal of efficiency), Dow Chemical Co. issued a press release announcing an "Energy Plan for America" calling for four major actions, the first of which is to "encourage aggressive efficiency and conservation."

KPMG study of 700 mergers: See (Collins, 2001). Also see James Surowecki's discussion of the KPMG study in "The Financial Page," *The New Yorker*, 9 June 2008 and 16 June 2008.

- Ayres, Robert U. "Energy Inefficiency in the U.S. Economy: A New Case for Conservatism." Laxenburg, Austria: International Institute for Applied Systems Analysis, 1989.

- Ayres, Robert U., Leslie W. Ayres, and Benjamin Warr. "Exergy, Power, and Work in the U.S. Economy, 1900–1998." *Energy* 28, no. 3 (2003): 219–273.

- Ayres, Robert U., Mark Narkus-Kramer, and Andrea L. Watson. "An Analysis of Resource Recovery and Waste Reduction Using SEAS." Washington, DC: International Research and Technology Corporation, 1976.

- Bridges, Jack. *Understanding the National Energy Dilemma (1973).* Washington, DC: United States Congress Joint Committee on Atomic Energy, 1973.

- Carnahan, Walter, Kenneth W. Ford, Andrea Prosperetti, Gene I. Rochlin, Arthur H. Rosenfeld, Marc H. Ross, Joseph E. Rothberg, George M. Seidel, and Robert H. Socolow. "Efficient Use of Energy: A Physics Perspective." New York: American Physical Society, 1975.

- Collins, Jim. "The Misguided Use of Acquisitions." In *Good to Great: Why Some Companies Make the Leap...and Others Don't.* New York: HarperCollins Business, 2001.

- Ehrhardt, Karen, and John A. Laitner. "The Size of the U.S. Energy Efficiency Market: Generating a More Complete Picture." Washington, DC: American Council for an Energy-Efficient Economy, 2008.

- Ferranti, Adele, Miriam Pye, Gary Davidson, and Dana Levy. "Encouraging P2 and E2 in New York." *Clearwaters* 30 (Spring 2000).

- Hawk, David. "Optimizing Savings Through a Steam Systems Approach." Massachusetts Energy Efficiency Partnership, 2006.

- Nakicenovic, Nebojsa, Paul V. Gilli, and Rainer Kurz. "Regional and Global Exergy and Energy Efficiencies." *Energy—The International Journal* 21 (1996): 223–237.

- Nelson, Kenneth E. "Dow's Energy/WRAP Contest: A 12-Year Energy and Waste Reduction Success Story." Houston, Texas: Industrial Energy Technology Conference, 1993.

5. The Future of Electric Power

PURPA: Public Utility Regulatory Policies Act, U.S. Code Sections 2601–2645. The act addresses cogeneration and small power production in Title 16, Chapter 12. PURPA was amended by the Energy Policy Act of 2005, Sections 1251–1254.

No credit for renewables in the "avoided cost" price: See (Kubiszewski, 2006).

Distribution monopolies and savings equivalent to $700 billion: See (Ayres, Turton, and Casten, 2007).

Power plants and air pollution: See (American Lung Association, 2009).

Bonanza and Cliffside power plant emissions rulings: See (www.sourcewatch.org, 2008).

Duke Energy's Cliffside plant: On its own web site (www. duke-energy.com), Duke noted, "Duke Energy has made significant improvements to reduce emissions from the company's coal-fired plants." It made no reference to carbon dioxide, the principal greenhouse gas, which its requested expansion would greatly increase.

Edison "patrolled" signs: The signs were found about 2 miles from the San Andreas Fault, at a point where the power line crosses a fire road in the Angeles National Forest south of Leona Valley, California.

Capital cost of new central power plants vs. decentralized generation: See (Casten and Collins, 2006).

Cabot Corporation recycling rebuffed in Louisiana: See (Ayres, Turton, and Casten, 2007).

James Hansen's letter to Duke Energy CEO Jim Rogers: Excerpted in "Cliffside, Coal, and Global Warming," www.nc.sierra-club.org.

- Ayres, Robert U., Hal Turton, and Tom Casten. "Energy Efficiency, Sustainability, and Economic Growth." *Energy* 32 (2007): 634–648.

- Casten, Thomas R., and Marty Collins. "WADE DE Economic Model." In *World Survey of Decentralized Electricity*.

Edinburgh, Scotland: The World Alliance for Decentralized Energy, 2006.

- Kubiszewski, Ida, et al. "Public Utility Regulatory Policies Act of 1978, United States (PURPA)." In *Encyclopedia of Earth*, edited by C. J. Cleveland. Washington, DC: Environmental Information Coalition, National Council for Science and the Environment, 2006.

6. Liquid Fuels: The Hard Reality

Ethanol studies: See, for example, (Williams, et al., 1994; Hammerschlag, 2006; Natural Resources Defense Council [NRDC], 2006; Jones, 2007).

Growing crops for ethanol vs. converting cropland to forest: See (Righelato and Spracklen, 2007).

Ethanol or methanol from cellulose: Whereas ethanol from corn can be expected to produce (at most) about 15 billion gallons a year in the United States by 2012 (around 6 percent of projected gasoline demand), the theoretical potential for ethanol from woody plants (and municipal refuse) is at least ten times as much, or 150 billion gallons a year, according to the Natural Resources Defense Council (NRDC). For further discussion, see www.informit.com/register.

Airplane fuel consumption: See (Murty, 2000).

Algae-based jet fuel: See (Gross, 2008)and (United States Energy Information Agency, 1998).

Dreamliner fuel use: See (Boeing Commercial Airplanes, 2008) and www.boeing.com/commercial/787family/background.html.

- Gross, Michael. "Algal Biofuel Hopes." *Current Biology* 18, no. 2 (2008).
- Hammerschlag, Roel. "Ethanol's Energy Return on Investment: A Survey of the Literature 1990." *Environmental Science & Technology* 40, no. 6 (2006): 1744–1750.
- Jones, Les. *Energy Return on Investment (EROI) 2007*. Cited 17 November 2007. Available from www.lesjones.com/posts/003223.shtml.

- Murty, Katta G. "Greenhouse Gas Pollution in the Stratosphere Due to Increasing Airplane Traffic, Effects on the Environment." Ann Arbor, Michigan: Department of Industrial and Operations Engineering, University of Michigan, 2000.
- Natural Resources Defense Council (NRDC). "Ethanol: Energy Well Spent: A Survey of Studies Published Since 1990." Natural Resources Defense Council, 2006.
- Righelato, Renton, and Dominick Spracklen. "Carbon Mitigation by Biofuels or by Saving and Restoring Forests?" *Science* 317 (2007): 902.
- United States Energy Information Agency. *Manufacturers Energy Consumption Survey 1998* [PDF or Lotus 123]. United States Energy Information Agency 1998. Cited 2002. Available from www.eia.doe.gov/emeu/mecs.
- Williams, Robert H., Eric D. Larson, Ryan E. Katofsky, and Jeff Chen. "Methanol and Hydrogen from Biomass for Transportation." Paper read at Biomass Resources: A Means to Sustainable Development, in Bangalore, India, 3–7 October 1994.

7. Vehicles: The End of the Affair

Domination of cities by cars: The American love affair with cars began to be seriously questioned in the 1960s and 1970s with the publication of Ralph Nader's *Unsafe at Any Speed* and with the emerging environmental movement marked by the first Earth Day in 1970. See (Ayres, 1970).

U.S. car crashes and fatalities: See (U.S. Department of Transportation Fatality Analysis Reporting System, 2007).

Energy efficiency of bicycles: Johns Hopkins University engineers recently measured the heat generated by friction as the drive chain of a bicycle moved through the sprockets under varying conditions. The chain drive posted an energy efficiency ranging from 81 to 98.6 percent (Johns Hopkins University, 1999).

Electric cars: The future of all-electric cars depends on two developments. One is the much lighter body made of composite fiber-reinforced polymers, together with aluminum alloys. Such a body can cut the weight of the vehicle—and, with it, the weight of the power plant needed—by half or more (Lovins, 1996). The other big problem for mass production of all-electric cars is mass production of high-performance batteries, almost certainly the rechargeable lithium-ion type that is now widely used for laptop computers and other such devices. However, some lingering safety concerns arise from the fact that lithium metal is extremely combustible, and electrical malfunctions have produced a few fires. Also, current output of lithium is small, and current production comes from a few dry salt beds in Bolivia, Argentina, and Chile. However, in the long run, lithium can probably be obtained from sea water at costs not much greater than the current price (Yaksic Beckdorf, and Tilton, 2008). However, the all-electric vehicle won't likely command a significant market share for at least the next two decades. Hybrids are a much more practical solution for the near term. See www. informit.com/ register.

Chinese e-bikes and batteries: See (Weinert, Burke, and Wei, 2007).

Car sharing: See (Bryner, 2008; Cervero, 2003). Also see The Car Sharing Network, www.carsharing.net and www.zipcar.com.

- Ayres, Edward H. *What's Good for GM*. Nashville: Aurora, 1970.
- Bryner, Jeanna. "Car Sharing Skyrockets As Gas Prices Soar." *US News and World Report*, 11 July 2008.
- Cervero, Robert. "Car Sharing Spurring Travel Changes." Berkeley, California: U.C. Berkeley Institute of Urban and Regional Development, 2003.
- Johns Hopkins University. "Wheel Power Probe Shows Bicycles Waste Little Energy." *Johns Hopkins Gazette*, 30 August 1999.
- Lovins, Amory B. "Hypercars: The Next Industrial Revolution." Paper read at 13th International Electric Vehicle Symposium (EVS 13), in Osaka, Japan, 14 October 1996.
- Weinert, Jonathan, Andrew Burke, and Xuezhe Wei. "Lead-Acid and Lithium-Ion Batteries for the Chinese Electric Bike Market and Implications on Future Technology Advancement." *Journal of Power Sources* 172, no. 2 (2007): 938–945.
- Yaksic Beckdorf, Andres, and John E. Tilton. "Using the Cumulative Availability Curve to Assess the Threat of Mineral

Depletion: The Case of Lithium." MS, Pontificio Universidad
Catolica de Chile, Santiago, Chile, 2008.

8. Preparing Cities for the Perfect Storm

Impacts of sea-level rise in California: See (California Environmental Protection Agency).

Grand Forks flood: See (Ayres, 1999).

European passive houses: See European Commission, "Promotion of European Passive Houses (PEP) Report," 2008. The PEP project is funded by European Commission, Energy and Transport, under contract number EIE/04/030/SO7.39990. Also see (Reisinger, et al., 2002; Elswijk and Kaan, 2008).

Emissions per passenger-mile: cars, light rail, and bus rapid transit: See (Vincent and Jeram, 2006; Vincent and Walsh, 2003).

What we learned from Brazil: See (Goodman, Laube, and Schwenk, 2005/2006).

Bus rapid transit on six continents: Compared to metro rail, BRT is much less expensive—as little as one-twentieth of the cost. For further information, see www.informit.com/register and EMBARQ, at the World Resources Institute (WRI) (e-mail EMBARQ@WRI.org). Also see (Herro, 2006).

Worst-case sea-level rise: NASA's chief climate scientist, James Hansen, wrote in 2007: "I find it almost inconceivable that 'business as usual' climate change will not result in a rise in sea level measured in metres within a century... . It seems to me that scientists downplaying the dangers of climate change fare better when it comes to getting funding. I can vouch for that from my own experience. After I published a paper in 1981 that described the likely effects of fossil fuel use, the U.S. Department of Energy reversed a decision to fund my group's research, specifically criticizing aspects of that paper." Hansen is not alone in this outlook; Tony Payne, professor of glaciology at the University of Bristol and co-director of the U. K.'s Centre for Polar Observation and Modeling, told a 2005 Royal Society conference of Antarctic climate experts, "The melting of the ice contained in West Antarctica would lead to a sea-level rise of five to six metres around the world, sufficient to cause effects such as the inundation of much of the state of Florida." The comment was reported by Environmental News Service, 18 October 2005.

Updated consensus of climate science: See (Intergovernmental Panel on Climate Change, 2007).

Desertification: See reports cited on the web site of the United Nations Convention to Combat Desertification, www.unccd. int/convention/menu.php.

U.S. Mayors Climate Protection Agreement: Office of the Mayor, Seattle, WA; 2008. See www.seattle.gov./Mayor/Climate.

The disappearing Mississippi Delta: See (Louisiana Coastal Wetlands Conservation and Restoration Task Force, 2008).

Rising risks to New York City: Cynthia Rosenzweig and Vivien Gomitz, NASA Goddard Institute for Space Studies, Columbia University, and New York City Department of Environmental Protection—see (McGeehin, 2008).

Valmeyer relocation: Operation Fresh Start: Using Sustainable Technologies to Recover from Disaster, a project of the National Center for Appropriate Technology, 2006. See www.freshstart.ncat. org/case/valmeyer.htm.

- Ayres, Edward H. *God's Last Offer: Negotiating for a Sustainable Future*. New York: Four Wall Eight Windows/Basic Books, 1999.
- Elswijk, Marcel, and Henk Kaan. *European Embedding of Passive Houses*. PEP project, 2008. Cited 15 January 2009. Available from www.aee-intec.at/0uploads/dateien578.pdf.
- Goodman, Joseph, Melissa Laube, and Judith Schwenk. "Curitiba's Bus System Is Model for Rapid Transit." *Race, Poverty, and the Environment* (2005/2006): 75–76.
- Herro, Alana. "Bus Rapid Transit Systems Reduce Greenhouse Gas Emissions, Gain in Popularity." In *Eye on Earth*. Washington, DC: WorldWatch Institute, 2006.
- Intergovernmental Panel on Climate Change (IPCC). *Report of the Working Group III of the IPCC*. Cambridge, U.K.: Cambridge University Press, 2007.

- Louisiana Coastal Wetlands Conservation and Restoration Task Force. "Standing Ground Against Advancing Waters Acre by Acre, CWPPRA Projects Beat Back Coastal Demise." *Water Marks* (2008).

- McGeehin, John P., et al. "Abrupt Climate Change." Washington, DC: U.S. Climate Change Program, U.S. Geological Survey, National Oceanic and Atmospheric Administration, National Science Foundation; 2008.

- Reisinger, Dulle, Henao, and Pitterman. *VLEEM—Very Long Term Energy Environment Modelling.* Vienna, Austria: Verbundplan, 2002.

- Vincent, Bill, and Brian Walsh. "The Electric Rail Dilemma: Clean Transportation from Dirty Electricity?" Washington, DC: Breakthrough Technologies Institute, 2003.

- Vincent, William, and Lisa Callaghan Jeram. "The Potential for Bus Rapid Transit to Reduce Transportation-Related CO_2 Emissions." *Journal of Public Transportation* (BRT Special Edition) (2006): 219–237.

9. The Water-Energy Connection

Falling water tables: See (Brown, 2006; Wilkinson, 2008).

Ethanol, energy, and water: See (National Research Council National Academy of Sciences, 2007). The report warns: "If projected increases in the use of corn for ethanol occur, the harm to water quality could be considerable, and water supply problems at the regional and local levels could arise.... The quality of groundwater, rivers, and coastal and offshore waters could be impacted by increased fertilizer and pesticide use for biofuels."

U.S. water consumption for irrigation: See (Maupin and Barber, 2005; (Abt, 1997).

U.S. water use for power plant cooling: See (Veil, 2007).

Desertification: National reports can be found by consulting the United Nations Convention to Combat Desertification (UNCCD) web site: www.unccd.int/convention/menu.php.

Energy cost of pumping in California: See (Trask, et al., 2005; Davis, 2005).

Pumping water to the North of China: See (SPG Media, 2009).

- Abt, Clark C. "China's Sustainable Growth Maximized by Avoiding Agricultural and Energy Shortages with Renewable Energy Resources for Farming, Irrigation, Transport, and Communications." Paper read at International Conference on China's Economy with Moderately Rapid and Stable Growth, in Guanxi Province, China, 2–4 September 1997.

- Brown, Lester R. "Water Tables Falling and Rivers Running Dry." In *Plan B 2.0: Rescuing a Planet under Stress and a Civilization in Trouble*, edited by L. R. Brown. New York: W. W. Norton and Co., 2006.

- California Environmental Protection Agency. DRAFT 2009 Climate Action Team Biennial Report to the Governor and Legislature, 1 April 2009.

- Davis, Martha. "Water-Energy Nexus." Sacramento, California: Inland Empire Utilities Agency (IEUA), 2005.

- Maupin, Molly A., and Nancy L. Barber. "Estimated Withdrawals from Principal Aquifers in the U.S. in 2000." Washington, DC: United States Geological Survey (USGS), 2005.

- National Research Council National Academy of Sciences. "Water Implications of Biofuels Production in the United States." Washington, DC: National Academy Press, 2007.

- SPG Media Ltd. *South-to-North Water Diversion Project.* Cited April 2009. Available from www.water-technology.net/projects/south_north/.

- Trask, Matt, Ricardo Amon, Shahid Chaudry, Thomas S. Crooks, Marilyn Davin, Joe O'Hagen, Pramod Kulkarni, Kae Lewis, Laurie Park, Paul Roggensack, Monica Rudman, Lorraine White, and Zhiqin Zhang. "California's Water–Energy Relationship." California Energy Commission, 2005.

- Veil, John A. "Use of Reclaimed Water for Power Plant Cooling." Chicago: Argonne National Laboratory (ANL), 2007.

- U.S. House of Representatives. Committee on Science and Technology, Subcommittee on Energy and the Environment. *Testimony of Robert Wilkinson, Ph.D.* 14 May 2008.

11. Implications for Business Management

Primary Energy's 900 megawatts: See (Downes, 2009).

Tom Casten and Recycled Energy Development: "Who Is Recycled Energy Development," at the RED web site: www. recycled-energy.com/.

Bill Gates and private investment in energy recycling: See Peter Robison, "Gates, Harvard Join a Record Energy-Recycling Fund," www.bloomberg.com/apps/news?pid=newsarchive &sid=aZoPAVvD_LNo.

Business alliances with nonprofit organizations (NPOs): See (Rondinelli and London, 2001).

Unsustainability of the modern economy: Humans currently consume 20 percent more natural resources than Earth can produce, reported the global conservation organization WWF in 2004. The report, based on the ecological footprint index, found that energy consumption has been the fastest-growing component of the index over the previous 40 years, increasing by 700 percent. Four years later, in 2008, the Global Footprint Network reported that the "over-shoot" appeared to be accelerating, and that "Humans now require the resources of 1.4 planets." Some scientists dispute the ecological footprint theory, arguing that humans can (and do) increase the carrying capacity of their environment to meet their needs—for example, by developing renewable energies. But even if true, that can't eliminate the deficit during the period of the transition bridge. Also see (Wackernagel and Rees, 1997; Boulding, 1966; Dietz, Rosa, and York, 2007).

- Boulding, Kenneth E. "The Economics of the Coming Spaceship Earth." In *Environmental Quality in a Growing Economy: Essays from the Sixth RFF Forum*, edited by H. Jarrett. Baltimore: Johns Hopkins University Press, 1966.

- Dietz, Thomas, Eugene A. Rosa, and Richard York. "Driving the Human Ecological Footprint." *Ecological Economics* 20, no. 1 (2007): 3–24.

- Downes, Brennan. "Potential of Energy Recycling and CHP in the U.S. Steel Industry." *Cogeneraton & On-Site Power* 10, no. 1 (2009).

- Rondinelli, Dennis A., and Ted London. "Partnering for Sustainability: Managing Nonprofit Organization–Corporate Environmental Alliances. Aspen Institute, 2001.

- Wackernagel, Mathis, and William E. Rees. "Perceptual and Structural Barriers to Investing in Natural Capital: Economics from an Ecological Footprint Perspective." *Ecological Economics* 20, no. 1 (1997): 3–24.

12. How Much, How Fast?

Subsidies for the oil industry and the price of gasoline: See (Harrje, Bricker, and Kallio, 1998; United States Energy Information Administration, 2007).

Summary of potential for the energy-transition bridge: See notes for Chapter 3.

- Harrje, Evan, Amy Bricker, and Karmen Kallio. "The Real Price of Gasoline," edited by M. Briscoe. Washington, DC: International Center for Technology Assessment, 1998.

- United States Energy Information Administration. "Federal Financial Interventions and Subsidies in Energy Markets, 2007." Washington, DC: United States Energy Information Agency (EIA), 2007.

INDEX